The Handbook of Graphene Electrochemistry

Dale A. C. Brownson · Craig E. Banks

The Handbook of Graphene Electrochemistry

 Springer

Dale A. C. Brownson
Craig E. Banks
Faculty of Science and Engineering, School
   of Science and the Environment, Division
   of Chemistry and Environmental Science
Manchester Metropolitan University
Manchester
UK

ISBN 978-1-4471-6427-2   ISBN 978-1-4471-6428-9   (eBook)
DOI 10.1007/978-1-4471-6428-9
Springer London Heidelberg New York Dordrecht

Library of Congress Control Number: 2014935967

Printed on acid-free paper

Springer is part of Springer Science+Business Media (www.springer.com)

My best moment? I have a lot of good moments but the one I prefer is when I kicked the hooligan.

Eric Cantona

Preface

Graphene, a one-atom thick individual planar carbon layer, has been reported to possess a range of unique and exclusive properties and is consequently being explored in a plethora of scientific disciplines. Although theoretically graphene has been scientifically investigated since the 1940s and was known to exist since the 1960s, the recent burst of interest can be correlated with work by Geim and Novoselov in 2004/2005, who reported the so-called "scotch tape method" for the production of graphene in addition to identifying its unique electronic properties. Thereafter, the 2010 Nobel Prize in Physics was awarded jointly to Geim and Novoselov *"for groundbreaking experiments regarding the two-dimensional material graphene"*. As a result there is a global pursuit to find new 'industrial scale' methodologies for the facile fabrication of pristine graphene and other graphene family members. Furthermore, graphene and related structures are being extensively incorporated into an ever diversifying range of applications across many areas in the search for greatly improved device performance.

One area which receives significant interest is the field of electrochemistry where graphene has been reported to be beneficial in various applications ranging from sensing through to energy storage and generation and carbon-based molecular electronics.

This handbook aims to provide readers with a fundamental introduction into electrochemistry, allowing one to be able to design and interpret experiments utilising graphene. Subsequent chapters consider current literature reports regarding fabricating graphene, utilising graphene as an electrochemical sensor and its impact on energy storage and production. Due to the range of currently available graphenes and those that will likely be fabricated in the near future along with the wide applications of graphene in electrochemistry, the area is truly fascinating.

March 2014

Dale A. C. Brownson
Craig E. Banks

Contents

Abbreviations

k°	Standard electrochemical rate constant (for a heterogeneous electron transfer reaction)
ΔE_p	Peak-to-peak separation
AFM	Atomic force microscopy
ASV	Anodic stripping voltammetry
BDD	Boron doped diamond
BPPG	Basal plane pyrolytic graphite
CRM	Certified reference material
CNT	Carbon nanotube
CVD	Chemical vapour deposition
CV	Cyclic voltammetry (or cyclic voltammogram)
DFT	Density functional theory
DPV	Differential pulse voltammetry
DOS	Density of electronic states
EDLC	Electrochemical double-layer capacitance
EPPG	Edge plane pyrolytic graphite
FWHM	Full width at half maximum
GC	Glassy carbon
GCP	Graphene-cellulose membrane
GNR	Graphene nano ribbon
GNS	Graphene nano sheet
GO	Graphene oxide
HOPG	Highly ordered pyrolytic graphite
IUPAC	International union of pure and applied chemistry
LOD	Limit of detection
LOQ	Limit of quantification
MFC	Microbial fuel cell
MWCNT	Multi-walled carbon nanotube
NADH	β-nicotinamide adenine dinucleotide
ORR	Oxygen reduction reaction
PBS	Phosphate buffer solution
RGS	Reduced graphene sheet
RSD	Relative standard deviation
SCE	Saturated calomel electrode

SECM	Scanning electrochemical cell microscopy
SEM	Scanning electron microscope
SHE	Standard hydrogen electrode
SPE	Screen-printed electrode
STM	Scanning tunnelling microscopy
SWCNT	Single-walled carbon nanotube
SWV	Square wave voltammetry
TEM	Transmission electron microscope
XPS	X-ray photoelectron spectroscopy

Chapter 1
Introduction to Graphene

In this chapter we first explore the fascinating story behind graphene's emergence onto the scientific horizon, thereafter focusing on the various methodologies for fabricating graphene before finally depicting the truly outstanding and exceptional properties of graphene to be reported in the literature, which have captured the imagination of scientists in a plethora of disciplines.

1.1 The Origins of Graphene

According to IUPAC, the suggested definition of graphene is: *a single carbon layer of the graphite structure, describing its nature by analogy to a polycyclic aromatic hydrocarbon of quasi infinite size* [1]. It is then noted by IUPAC that *previously, descriptions such as graphite layers, carbon layers or carbon sheets have been used for the term graphene. Because graphite designates that modification of the chemical element carbon, in which planar sheets of carbon atoms, each atom bound to three neighbours in a honeycomb-like structure, are stacked in a three-dimensional regular order, it is not correct to use for a single layer a term which includes the term graphite, which would imply a three-dimensional structure. The term graphene should be used only when the reactions, structural relations or other properties of individual layers are discussed* [1]; a conceptual depiction along with scanning electron microscope (SEM) and transmission electron microscope (TEM) images of the graphene structure are shown in Fig. 1.1.

1.1.1 Graphene: A Brief History

The exact history of graphene and how it appeared on the scientific horizon is fascinating. In theory, as an integral part of various three-dimensional materials, graphene has been studied since the 1940s [2–4]. In 1947 Philip Wallace wrote a pioneering paper concerning the electronic behaviour of graphite that sparked

D. A. C. Brownson and C. E. Banks, *The Handbook of Graphene Electrochemistry*, DOI: 10.1007/978-1-4471-6428-9_1, © Springer-Verlag London Ltd. 2014

Fig. 1.1 A conceptual model depicting the structure of graphene (**a**) and TEM (**b**)/ SEM (**c**) images of a single atomic layer of graphite, known as graphene. A high-resolution TEM (**d**) image is also shown, where the *white arrow* indicates the edge of the graphene sheet. Note, in reality, the graphene utilised in the majority of work deviates from that of 'true' graphene. **b** and **d** are reproduced from Ref. [8] with permission from The Royal Society of Chemistry and **c** is reproduced from Ref. [9] with permission from Elsevier

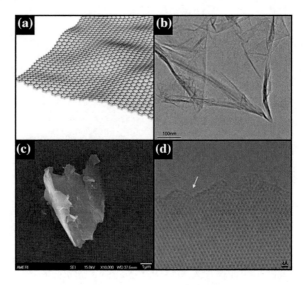

interest into the exploration of graphene [3], however it was not until the recent work of Novoselov et al. [5, 6] and Zhang et al. [7] that interest in graphene escalated due to reports of its unique properties [8].

In 2004 Novoselov et al. reported the development of a simple, yet time consuming methodology in which one could produce and observe microscopic few-layer graphene crystals on silicon wafers (silicon dioxide on silicon) [5]. Subsequently, this technique has been copied globally as a protocol to produce large area single layer graphene samples, allowing two-dimensional transport studies to be performed [11]. Resultantly, in 2010 the Nobel Prize in Physics was awarded jointly to Geim and Novoselov for ground breaking experiments regarding the two-dimensional material graphene [12]. However, as highlighted by de Heer [11], there is a common misconception regarding the 2004 paper by Novoselov and Geim. In his letter to the Nobel Committee addressing such issues (a copy of which can be found at Appendix A), de Heer noted that the majority of scientific publications incorrectly cite the 2004 report as the paper that presented both the 'scotch tape method' and graphene's unique electronic properties to the world [11]. In fact such findings were not reported with regards to individual single layer graphene in 2004 [5], but actually in a 2005 paper by Novoselov and Geim [6, 11]. Furthermore, in reality, graphene had been identified and characterised as a two-dimensional-crystalline material in many reports prior to 2004 where ultra-thin graphitic films were observed and occasionally even monolayer graphene (see for example Refs. [8] and [13] for pertinent reviews) [11, 14].

Dryer et al. [13] have elegantly produced a detailed account of the synthesis and characterisation of graphene; Fig. 1.2 shows a timeline representing the history of the preparation, isolation and characterisation of graphene as given in Ref. [13]. Of note is that in 1962, H.-P. Boehm, who coined the name graphene in 1986 [15–17], reported his observations of graphene and demonstrated beyond reasonable doubt

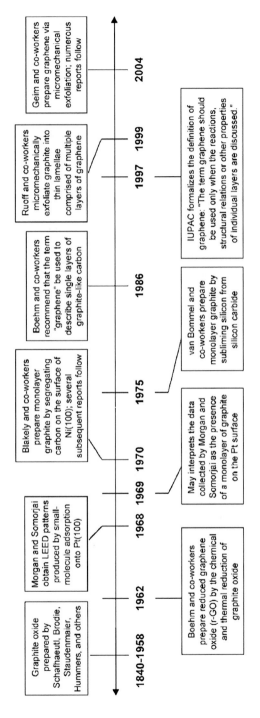

Fig. 1.2 A timeline of selected events in the history of graphene for its preparation, isolation and characterisation. Reproduced from Ref. [12] with permission from Wiley

the existence of freestanding graphene [15–17]. Dryer et al. point out that in the 1962 report, Boehm et al. isolated reduced graphene oxide with heteroatomic contamination rather than pristine graphene [18], where ultimately as a result of this the electrical conductance is significantly lower for this material than for pristine graphene prepared by the 'scotch tape method' [17–20]. Of historical significance is that, excluding the work of Boehm et al. reports of graphene prior to 2004 were merely observational and failed to describe any of graphene's distinguishing properties [8, 13]. Thus, the 2005 report by Novoselov and Geim can be considered as the first to report both the isolation of 'pristine' graphene (i.e. single layer graphene without heteroatomic contamination) and its unique properties to the world; which in doing so sparked the graphene gold rush and brought new and exciting physics to light [8, 13].

Since the pioneering reports of 2004/2005 many other unique properties have been assigned to graphene (see Sect. 1.3) and a significant array of other methodologies has been reported regarding its fabrication (see Sect. 1.2) [21–29]. Graphene has truly captured the imagination of scientists from around the globe and is now an extensive and vibrant area of research, where its utilisation has resulted in an improved understanding of fundamental factors in addition to significantly enhanced device performance in a wide range of scientific fields.

1.1.2 Graphene: Meet the Family

Pristine graphene is a two-dimensional sp^2 bonded carbon nanostructure [2, 30]. Most significantly however, graphene is a key derivative of carbon and originates from a large family of fullerene nanomaterials where it is the essential 'building block' for many of the allotropic dimensionalities that have significant and widespread use as electrode materials [2, 30].

The constituent atoms of graphite, fullerenes and graphene share the same basic structural arrangement in that each structure begins with six carbon atoms which are tightly bound together (chemically, with a separation of ~ 0.142 nm) in the shape of a regular hexagonal lattice [31]. At the next level of organisation graphene is widely considered as the 'mother of all graphitic forms', where, as depicted in Fig. 1.3, in addition to existing in its planar state a singular graphene sheet can be 'wrapped' into a zero-dimensional spherical C_{60} buckyball, 'rolled' into a one-dimensional carbon nanotube (CNT) (further categorised into single- or multi-walled depending on the number of graphene layers present (SWCNTs/MWCNTs respectively)), or multiple graphene sheets can be 'stacked' into three-dimensional graphite (generally consisting of ≥ 8 graphene layers, see below for further details); the stacked graphene sheets/planes in graphite are separated by a distance of 0.335 nm and are held together by weak, attractive intermolecular forces [31].

It is important to note that the graphene structure itself (that is, in its standard pristine form, a single layer/sheet of the carbon structure) is often referred to as a graphene nano sheet (GNS) which implies a large, scalable graphene sheet of

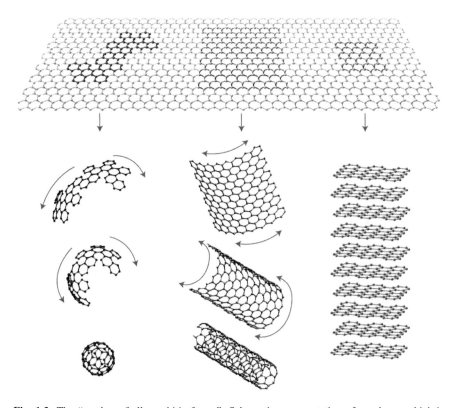

Fig. 1.3 The "mother of all graphitic forms". Schematic representation of graphene, which is the fundamental starting material for a variety of fullerene materials; C_{60} (buckyballs) (*bottom row* at the *left*), CNTs (*bottom row* in the *centre*), and graphite (*bottom row* at the *right*). Reproduced from Ref. [2] with permission from the Nature Publishing Group

'quasi-infinite size'; however, variations in the graphene structure do exist. Graphene nano ribbons (GNRs) are strips of graphene that possess an ultra-thin width (<50 nm) (note that other various shapes also exist) [32], whereas graphene nano-platelets are a further variation (often referred to as double-, few-, or multi-layered graphene sheets) which are characterised by stacks consisting of between 2 and 7 graphene sheets and thus should not be considered as graphene (which implies 1 individual carbon layer) or graphite (which implies a structure of 8 or more graphene layers) but as an intermediate phase with distinct properties that vary accordingly with increasing layer numbers until the graphite structure is achieved [33–36]; this intermediate phase of graphene/graphite is known as quasi-graphene [37, 38], however given the varying properties it is wise to state the number of graphene layers when working within this subgroup. Increasing the number of graphene layers past 8 results in negligible alterations in terms of the evolution of the electronic structure and various other properties of graphene and such structures are thus to be considered as graphite (as determined by Raman spectroscopy and scanning electrochemical cell microscopy (SECM)) [33–36].

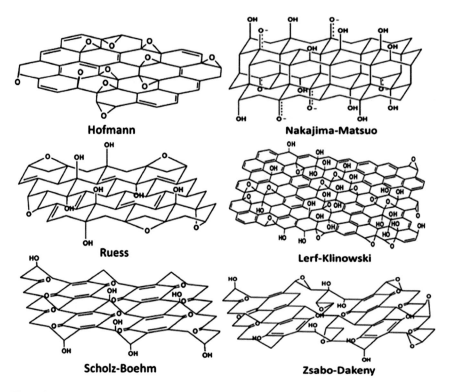

Fig. 1.4 Proposed configurations for GO when synthesised via varying routes. Reproduced from Ref. [36, 37] with permission from The Royal Society of Chemistry

Finally, another form of graphene that is commonly employed across the literature is that of graphene oxide (GO). GO consists of graphene that has been oxidised within the employed fabrication process or spontaneously by contact with air, however, this form is usually chemically or electrochemically reduced before use (see Sect. 1.2.3) [41]. Depending on the fabrication approach utilised to synthesise GO its structure varies significantly in terms of the presence of specific oxygenated species and their given quantities; Fig. 1.4 depicts the proposed structure of GO as produced through varied fabrication methodologies. It is important to bear in mind that different oxygenated species present on various GOs will significantly influence the observed electrochemical response.

1.2 Fabricating Graphene

Preparative methods of graphene are currently a heavily researched and important issue. The search for a methodology that can reproducibly generate high quality monolayer graphene sheets with large surface areas and large production volumes is greatly sought-after. Consequently, several physical and chemical methods exist for the production of graphene, which include the mechanical or chemical exfoliation of graphite, unrolling of CNTs (either through electrochemical, chemical or physical methods), Chemical Vapour Deposition (CVD) or epitaxial growth, reduction of GO and many other organic synthetic protocols [18, 30, 42–44]; researchers are currently exploring new and innovative approaches for the facile fabrication of graphene. It is important to note however that each method has innate advantages and disadvantages in terms of the resultant quality (properties), quantity and thus electrochemical applicability of the graphene produced and that there is presently no single method that exists for the production of graphene sheets that are suitable for all potential applications [30, 44]. Table 1.1 gives a direct comparison of selected graphene fabrication methodologies with their inherent advantages and disadvantages.

1.2.1 Mechanical Exfoliation

Among the methods stated above, dry mechanical exfoliation remains one of the most popular and successful methods for producing single- or few-layers of graphene [45]. This approach is known as the so-called "scotch tape method". As highlighted in Fig. 1.5, D.I.Y. graphene is possible given that this method is relatively simple, the process involves cleaving a sample of graphite (usually HOPG) with a cellophane-based adhesive tape [5]. The number of graphene layers formed can be controlled to a limited degree via the number of repeated peeling steps performed prior to the flakes then being transferred to appropriate surfaces for further study; note however that as depicted in Figs. 1.5 and 1.6a, when present, single layer graphene crystals are usually integrated with few- and multi-layer graphene crystals and it is a difficult and time consuming process to separate out these individual graphene sheets. Nonetheless, this method is ideally suited for the investigation of graphene's physical properties given that it does allow the low cost isolation of single graphene sheets that are of high quality [5, 43], however, disadvantages including poor reproducibility, low-yield and the labour intensive processes required result in it being difficult to scale this process to mass production and have thus lead to this method being used predominantly only for fundamental studies [43]. Moreover, the process generally yields graphene flakes of small sizes, although graphene flakes with sizes of up to 1 mm have been obtained; see Fig. 1.6b. A further disadvantage of this process is the possible damage (disrupting the basal surface, viz the generation of edge plane like-sites/defects) and contamination of the graphene samples, particularly from the adhesive utilised in the cellophane-based tape, which renders this method less appealing for the electrochemical investigation of pristine graphene.

Table 1.1 Comparison of various graphene fabrication methodologies commonly utilised to obtain graphene for electrochemical studies

Fabrication method	Graphene precursor	Operating conditions	Advantages	Disadvantages	Application implications	References
Mechanical exfoliation	HOPG	Scotch-tape	Direct, simple, high structural and electronic quality, low cost	Delicate and time-consuming (hours), low yields, poor reproducibility, possible contamination of sample from the adhesive tape utilised	Fundamental research. High quality single layer graphene sheets obtained with little lattice defect density and domain sizes ranging from 500 Å up to 10 μm	[4, 40, 42]
Chemical exfoliation	Graphite	Dispersion and exfoliation of graphite in organic solvents or through the use of surfactant complexes	Direct, simple, large-scale production, low-cost, high yield, practicability of sample handling (liquid suspension)	Time-consuming (hours), impure, possible contamination of sample from surfactant or solvents utilised	General graphene research for modified substrates. Often multiple layered graphene incorporated with structural defects originating from the fabrication process with domain size ranging from 500 up to 1500 Å	[40–42, 44, 45]
Reduction of GO	Graphite	Graphite exfoliation and oxidation, subsequent reduction of exfoliated graphite oxide	Facile scalability, high yields, low cost, excellent processability, practicability of sample handling (liquid suspension)	Indirect, large number of structural defects, disruption of the electronic structure of graphene owing to impurities, reduction to graphene is often not complete	General graphene research for modified substrates. Often multiple layered graphene incorporated with structural defects originating from the fabrication process with domain size ranging from 500 up to 1500 Å	[29, 40–42]

(continued)

Table 1.1 (continued)

Fabrication method	Graphene precursor	Operating conditions	Advantages	Disadvantages	Application implications	References
CVD epitaxial growth	Hydrocarbon gas (primarily)	CVD under variable temperatures and pressures (see Table 1.2)	Large-scale production, high qualities, uniform films, tailoring of graphene quality possible (see Table 1.2)	High temperature requirements, high cost, complicated process, variable yields	Fundamental and basic research. High quality single layer graphene sheets obtained with little lattice defect density, however, graphene can be tailored to contain specific defects and impurities where these are required for beneficial implementation in given devices. Layer thickness and domain sizes are thus variable (see Table 1.2)	[42, 46, 47]

Reproduced from Ref. [43] with permission from The Royal Society of Chemistry
Abbreviation: *CVD* chemical vapour deposition; *GO* graphene oxide; *HOPG* highly ordered pyrolytic graphite

D.I.Y. Graphene

1 Work in a clean environment; stray dirt or hair plays havoc with graphene samples.

2 Prepare a wafer of oxidized silicon, which helps you see graphene layers under a microscope. To smooth out the surface to accept the graphene and to clean it thoroughly, apply a mix of hydrochloric acid and hydrogen peroxide.

3 Attach a graphite flake to about six inches of plastic sticky tape with tweezers. Fold the tape at a 45-degree angle right next to the flake, so that you sandwich it between the sticky sides. Press it down gingerly and peel the tape apart slowly enough so that you can watch the graphite cleaving smoothly in two.

4 Repeat the third step about 10 times. This procedure gets harder to do the more folds you make.

5 Carefully lay the cleaved graphite sample that remains stuck to the tape onto the silicon. Using plastic tongs, gently press out any air between the tape and sample. Pass the tongs lightly but firmly over the sample for 10 minutes. With the tongs, keep the wafer planted on the surface while slowly peeling off the tape. This step should take 30 to 60 seconds to minimize shredding of any graphene you have created.

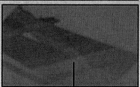

Graphene

6 Place the wafer under a microscope fitted with a 50× or 100× objective lens. You should see plenty of graphite debris: large, shiny chunks of all kinds of shapes and colors (*upper image*) and, if you're lucky, graphene: highly transparent, crystalline shapes having little color compared with the rest of the wafer (*lower image*). The upper sample is magnified 115×; the lower 200×.

—*JR Minkel, online news reporter*

Fig. 1.5 "D.I.Y. graphene: how to make one-atom-thick carbon layers with sticky tape". Reproduced with permission from Ref. [30]. Copyright J. R. Minkel, Scientific America, Inc. 2008

1.2.2 Chemical Exfoliation

An alternative preparative method that is commonly utilised owing to the ease of production, high-yield and relative low cost is the chemical exfoliation of graphite [43]. This includes ultrasound in both solution and intercalation steps, usually prior to

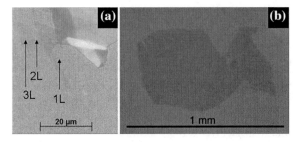

Fig. 1.6 **a** Optical microscopy image of single-, double- and triple-layer graphene (labelled as 1L, 2L and 3L respectively) on Si with a 300 nm SiO_2 over-layer. Reproduced from Ref. [48] with permission from Elsevier. **b** Monolayer graphene, produced by mechanical exfoliation, on a Si/SiO$_2$ wafer. This is a large sample with a length of 1 mm. Reproduced with permission from Ref. [49]. Copyright 2012 D. R. Cooper et al

the implementation of a centrifugation technique. For example, one ultrasonication route entails the use of a water-surfactant solution, sodium cholate [48], which forms stable encapsulation layers on each side of the graphene sheets; graphite flakes are dispersed in the aqueous surfactant solution and transformed into monolayer graphene by the application of ultrasound, resulting in graphene-surfactant complexes having buoyant densities that vary with graphene thickness [48, 53]. Following sonication the obtained solutions undergo centrifugation, which results in a 'sorting' of the graphene and hence different fractions are observed meaning that graphite and multi-layer graphene are not inadvertently incorporated into the graphene samples, after which the upper part of the resultant supernatant contains single layers of graphene floating in the solution which are then transferred using a pipette and dropped onto the surface of choice for further study [48]. Note that graphene fabricated via this route is readily commercially available [53, 54]. This procedure is also possible without additives in many organic solvents that have a high affinity for graphite where ultrasonic agitation is used to supply the energy to cleave the graphene precursor [24]. The success of ultrasonic cleavage depends on the correct choice of solvents and surfactants as well as the sonication frequency, amplitude and time [47]. As with mechanical exfoliation, the quality of the obtained graphene is not always sufficient (structural damage to the graphene can occur during preparation owing to ultrasonication, which may result in the graphene possessing a high defect density) and additionally homogeneity of the number of graphene layers is often poor [43], thus graphitic impurities may remain. Note also that material produced via such means often contains remains from the exfoliating agents utilised. These impurities can significantly affect the observed electrochemical characteristics and performance of the 3.2.3 graphene sample (see later for further details—Sects. 3.2.2 and 3.2.3).

1.2.3 Reduction of GO

Another popular aqueous based synthetic route for the production of graphene utilises GO [30, 43]. GO is produced via *graphite* oxide which itself can be fabricated via various different routes. The Hummers method for example involves soaking graphite in a solution of sulphuric acid and potassium permanganate to produce graphite oxide [43, 55]. Stirring or sonication of the graphite oxide is then performed to obtain single layers of GO—this is achieved given that GO's functional groups render it hydrophilic, allowing it to be dispersed in water based solutions. Finally, GO is chemically, thermally or electrochemically reduced to yield graphene [42, 43]. The majority of graphene used in electrochemistry is produced through the reduction of GO (often referred to as 'reduced GO' or 'chemically modified graphene'), and it is important to note that graphene produced in this manner usually has abundant structural defects (edge plane like-sites/defects) [18, 56] and remaining functional groups which results in partially functionalised graphene (thus is not pristine graphene): this has implications with regards to contributory factors influencing the observed electrochemistry (see Sect. 3.2.5). This method has the advantages of being scalable, rapid and cost effective in addition to the beneficial handling versatility of the liquid suspension; [44] however (as stated above) reduction to graphene is often only partial, lattice defects and graphitic impurities can also remain after reduction and additional interferences may arise in this case from the presence of reducing agents.

1.2.4 Miscellaneous Fabrication

Recent developments have led to the commercial availability of 'pristine' graphene that is produced via a substrate-free gas-phase synthesis method [9, 27, 57]. This single-step technique involves sending an aerosol consisting of liquid ethanol droplets and argon gas directly into a microwave-generated argon plasma (at atmospheric-pressure), where over a time scale in the order of 10^{-1} s, ethanol droplets evaporate and dissociate in the plasma forming solid matter that through characterisation by Transmission Electron Microscopy (TEM) and Raman spectroscopy is confirmed to be clean and highly ordered graphene sheets that are similar in quality to the graphene obtained through the mechanical exfoliation of HOPG [9, 27, 57]. In the case of commercially available graphene, the fabricated graphene sheets are sonicated in ethanol to form a homogeneous suspension before being distributed by the supplier [57]. Production in this manner has proven that graphene can be created without the use of three-dimensional materials as precursors or supporting substrates, and has demonstrated the viability of the large-scale synthesis of graphene.

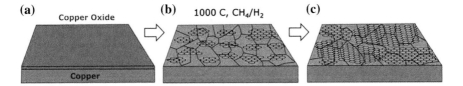

Fig. 1.7 Schematic illustrating the three main stages of graphene growth on copper by CVD: **a** copper foil with native oxide; **b** the exposure of the copper foil to CH_4/H_2 atmosphere at 1000 °C leading to the nucleation of graphene islands; **c** enlargement of the graphene flakes with different lattice orientations. Reproduced from Ref. [57] with permission from The Royal Society of Chemistry

1.2.5 CVD Fabrication

One of the most interesting fabrication approaches is the CVD growth of graphene. This method appears ideally suited for applications within electrochemistry with regards to the prevalence of uniform graphene sheets with high crystal quality and large surface areas, which are readily transferable and can be obtained at large manufacturing volumes [58, 59]. Additionally, CVD fabricated graphene is often supported on a (desirable and suitable) solid substrate and as such the positioning and orientation of the graphene can be precisely manipulated for specific purposes. This alleviates issues with regards to the controlled placement of solution based graphene sheets and in terms of the natural formation of graphite once the solvent is removed, as has been reported in some cases [46]. The underlying principle of CVD is to decompose a carbon feedstock with the help of heat in order to provide a source of carbon which can then rearrange to form sp^2 carbon species. This is usually accomplished over a catalyst [43]; for the growth of graphene, hydrocarbon gases are generally utilised as precursors and the most successful catalysts thus far are transition metal surfaces (namely nickel and copper) [49]. Figure 1.7 illustrates the three main stages of graphene growth over a copper catalyst via a CVD process.

Since the discovery of a uniform deposition of high-quality single layered graphene on copper there has been significant interest in the exploration of copper as a catalyst for the CVD growth of graphene [60]. It has been established that the most suitable catalysts for graphitic carbon formation are those transition metals that have a low affinity towards carbon but that are still able to stabilise carbon on their surfaces by forming weak bonds [60]. Interestingly, when compared to the alternative transition metals utilised in the CVD fabrication of graphene, copper has the lowest affinity to carbon (as reflected by the fact that it does not form any carbide phases) and has very low carbon solubility compared to Co and Ni (0.001–0.008 wt% at 1084 °C for Cu, 0.6 wt% for Ni at 1326 °C and 0.9 wt% for Co at 1320 °C) [60]. Copper's low reactivity with carbon can be attributed to the fact that it has a filled 3d-electron shell $\{[Ar]3d^{10}4s^1\}$, the most stable configuration (along with the half filling $3d^5$) because the electron distribution is symmetrical which minimises reciprocal repulsions [60]. As a result, Cu can form only

Fig. 1.8 SEM images of graphene on copper grown by CVD. **a** Graphene domain grown at 1035 °C on copper at an average growth rate of ∼6 μm/min. **b** Graphene nuclei formed during the initial stage of growth. **c** High-surface-energy graphene growth front shown by the *arrow* in (**a**). Reprinted with permission from Ref. [58]. Copyright 2011 American Chemical Society

soft bonds with carbon via charge transfer from the π electrons in the sp^2 hybridised carbon to the empty 4s states of copper [60]. Hence this peculiar combination of very low affinity between carbon and copper along with the ability to form intermediate soft bonds makes copper a true catalyst, as defined by textbooks, for graphitic carbon formation (whereas the $3d^7$ and $3d^8$ orbitals of Co and Ni are between the most unstable electronic configuration (Fe) and the most stable one (Cu)) [60]. Furthermore, note that often in the pursuit of obtaining large graphene domains upon transition metal catalysts, the pre-treatment of the said foils (i.e. annealing) has been found to be of vital importance: Refs. [46] and [60] provide detailed overviews and further insights for interested readers.

In recent groundbreaking work, single graphene crystals with dimensions of up to 0.5 mm were grown by low-pressure CVD in copper-foil enclosures using methane as a precursor [61]. Low-energy electron microscopy analysis showed that the large graphene domains had a single crystallographic orientation, with an occasional domain having two orientations [61]. The authors report that Raman spectroscopy revealed the graphene crystals to be uniform monolayer's with a low D-band intensity [61]. Scanning Electron Microscope (SEM) images of the fabricated graphene are presented in Fig. 1.8—this work was the first to report the growth of high quality large-grain-size *single* graphene crystals. However, bear in mind that recent studies of graphene produced by CVD on copper (and particularly on nickel) [46] showed that the majority of graphene produced is few- and multi-layered, in addition to being polycrystalline where its resultant mechanical strength is weakened (and its electronic properties altered) at the grain boundaries of the underlying substrate. These sites are the origin of surface defects in graphene (graphitic impurities) and thus in an electrochemical sense the degree of defect coverage will strongly influence the electrochemical properties of the graphene film [46]. One of the major advantages when utilising CVD in the fabrication of graphene is the variability in the graphene structure obtained at various conditions and hence the CVD process allows one to tailor the surface composition/structure [46], which will have inherent implications on its electrochemical performance—see Table 1.2 and Fig. 1.9 for example.

Furthermore, note that for the case of the CVD fabrication of CNTs, metallic impurities are commonly the 'hidden' origin of electrochemical activity for many analytes, which is inherent to the CVD fabrication process, where the amount of metallic impurities varies greatly between batches and hinders exploitation, for example in the fabrication of reliable CNT based sensor and energy devices [66]. It is clear that where graphene is fabricated via CVD appropriate control experiments will need to be performed in order to confirm the absence of such metallic impurities. Additionally, given the possible contribution of the underlying metal support/catalyst towards the observed electrochemistry at CVD grown graphene (where incomplete coverage of the graphene layer occurs—see Sects. 3.2.6), investigations towards the utilisation of non-metallic catalysts for graphene's CVD synthesis are also encouraged to overcome this potential issue. Alternatively, transfer of the fabricated graphene onto a more suitable insulating substrate is often necessary and is possible [46, 59].

1.2.6 Fabrication for Electrochemical Applications

Most of the interesting applications of graphene require growth of single-layer graphene onto a suitable substrate in addition to controllable coverage and favourable manipulation/positioning of the graphene, which is very difficult to control.

The more 'practical' solution-chemistry based approaches towards the fabrication of graphene are currently favoured for electrochemical applications because of their high yields and the flexibility of handing the graphene obtained from these processes. [45] In these cases the graphene is usually dispersed into a solvent which is then cast onto a suitable surface, where following evaporation the graphene left immobilised can then be used experimentally. This process, while facile, has inherent disadvantages such as surface instability, reproducibility and uncertainty issues in terms of the coverage and quality of the graphene remaining where deviation from true single layer 'pristine graphene' may exist (viz graphitic impurities). Consequently, in these instances post-application characterisation of the graphene is required for clarity [44]. Yet it is an effective way to explore the electrochemical properties of graphene.

Table 1.2 Comparison of variable CVD fabrication protocol/conditions utilised and the resultant variability in graphene quality obtained

Substrate/ catalyst	Temperature/°C	Gas reaction mixtures (precursors)	Growth time	Special conditions	Graphene grain size	Thickness of graphene layer	Graphene quality	References
Nickel	1,000 cooling rate ~10 °C s⁻¹	CH₄:H₂:Ar at 50:65:200 standard cubic centimetres per minute (sccm). Ambient pressure	7 min	Nickel thickness less than 300 nm deposited on Si/SiO₂ substrate. Prior annealing of nickel substrate	≤20 μm	~1–12 layers	Highly polycrystalline surface, small grain sizes and multilayered regions of graphene result in an extremely large degree of edge plane surface defects in the graphene film	[60]
Nickel	1,000 cooling rate of 100 °C min⁻¹	CH₄ at 10 sccm, H₂ at 1,400 sccm. Ambient pressure	5 min	Nickel thickness ~500 nm deposited on Si/SiO₂ substrate. Prior annealing of nickel substrate	3–20 μm	1, 2 and multi-layered graphene regions occupy up to 87 % of the film area and single layer coverage accounts for ~5–11 % of the overall film	Highly polycrystalline surface, small grain sizes and areas of few layered graphene islands result in a large degree of edge plane surface defects in the graphene film	[61]
Copper	800 cooling rate not specified	H₂/CH₄ at 5 sccm and partial pressure 0.39 Torr (Ar at 80 sccm, 1 Torr)	10 min	Copper foil (206 nm thick)	~10 μm	1, 2 and 3 layers	Few crystallographic orientations and edge plane defects present at the grain boundaries and at variable multiple layer graphene areas	[62]

(continued)

Table 1.2 (continued)

Substrate/ catalyst	Temperature/°C	Gas reaction mixtures (precursors)	Growth time	Special conditions	Graphene grain size	Thickness of graphene layer	Graphene quality	References
Copper	1,000 cooling rate 40–300 °C min^{-1}	H$_2$/CH$_4$ at 0.06 sccm and partial pressure 0.5 Torr	<3 min	Copper foil (25 μm thick)	10 μm	~95 % 1 monolayer	Few crystallographic orientations and few defects present at the grain boundaries with ~5 % being multiple layer graphene	[46]
Copper	~1,035 cooling rate not specified	CH$_4$ at a flow rate and partial pressure less than 1 sccm and 50 m Torr respectively	>1 h	Copper foil (25 μm thick), enclosure utilised	0.5 mm	1 monolayer	Single crystallographic orientation, high purity defect free single graphene crystals	[58]

Reproduced from Ref. [43] with permission from The Royal Society of Chemistry

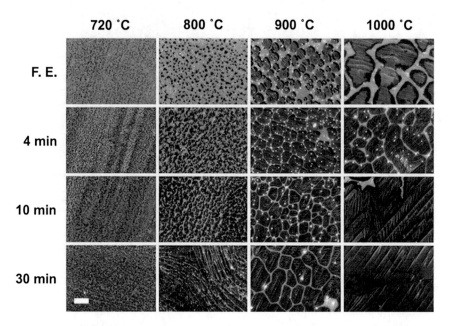

Fig. 1.9 High-resolution SEM images of graphene nuclei grown on Cu for different growth temperatures and times. These are identifiable as areas darker than the exposed Cu surface, which quickly oxidise in air after being taken out from the CVD growth system. Scale bar: 1 μm. Reprinted with permission from Ref. [59]. Copyright 2012 American Chemical Society

With future work in this field focusing on the use of single crystal substrates for the CVD growth of graphene, through careful control of the experimental conditions one can envisage high quality, contaminant free, single layered graphene crystals of bespoke sizes (or alternatively selectively impure graphene with customised properties) [46] being made commercially available within the near future. Such variability is beneficial for the electrochemical utilisation of graphene. It is however important to note; given that the structural characteristics and/or composition of graphene are likely to vary significantly depending on the fabrication route utilised, it is essential that any such fabricated graphene nanomaterial is thoroughly characterised prior to its implementation within electrochemistry to avoid potential misinterpretation of the experimental data. Additionally for convention, the way the graphene is fabricated should be referenced in its naming, for example, Staudenmaier thermally reduced GO, Hummers produced graphene and so on.

1.3 The Unique Properties of Graphene

Since the isolation of graphene and the reporting of its exceptional electronic properties in 2005 there has been a gold-rush in terms of exploring the full range of properties it has to offer. Table 1.3 summarises some of the reported astonishing

Table 1.3 Some reported properties of graphene

Property	Details	References
Optical transparency	97.7 %	[39, 64]
Electron mobility	200,000 cm^2 V^{-1} s^{-1}	[39, 64]
Thermal conductivity	5,000 Wm^{-1} K^{-1}	[39, 64]
Specific surface area	2,630 m^2 g^{-1}	[39, 64]
Breaking strength	42 N m^{-1}	[39, 64]
Elastic modulus	0.25 TPa	[39]

properties of graphene that have been determined to date. Graphene is reported in the media to be the "thinnest, most flexible and strongest material known".

Clearly graphene has captured the imagination of scientists and is now a hugely active area of research in a plethora of fields, none more so than in the field of electrochemistry which has reported many benefits in the areas of sensing through to energy storage and generation.

1.3.1 Electrochemically Important Properties

Carbon materials have been widely utilised in both analytical and industrial electrochemistry, where in many areas they have out-performed the traditional noble metals. This diversity and success stems largely from carbons structural polymorphism, chemical stability, low cost, wide potential windows, relatively inert electrochemistry, rich surface chemistry and electro-catalytic activities for a variety of redox reactions [30, 68].

When reviewing the essential characteristics of an electrode material for widespread applicability within electrochemistry graphene's 'theoretical advantage' becomes apparent. An essential characteristic of an electrode material is its surface area, which is important in applications such as energy storage, biocatalytic devices and sensors. Graphene has a theoretical surface area of 2630 m^2 g^{-1}, surpassing that of graphite (~ 10 m^2 g^{-1}), and is two times larger than that of CNTs (1315 m^2 g^{-1}) [69]. The electrical conductivity of graphene has been calculated to be ~ 64 mS cm^{-1}, which is approximately 60 times greater than that of SWCNTs [10, 70]. Furthermore, graphene's conductivity remains stable over a vast range of temperatures, encompassing stability at temperatures as low as liquid-helium, where such stability is essential for reliability within many applications [2]. More interestingly, graphene is distinguished from its counterparts by its unusual band structure, rendering the quasiparticles in it formally identical to the massless Dirac Fermions [30]. A further indication of graphene's extreme electronic quality is that it displays the half-integer quantum Hall effect, with the effective speed of light as its Fermi velocity, $v_F \approx 10^6$ m s^{-1}, which can be observed in graphene even at room temperature [71–73]. Resultantly, ultra high electron mobility has been achieved in suspended graphene [72, 74], where mobilities in excess of 200,000 cm^2 V^{-1} s^{-1} have been reported at room temperature. In comparison, the mobility of an electron in silicon is at its maximum at around 1,000 $cm^2 V^{-1} s^{-1}$, meaning the electron mobility is 200 times higher in graphene [30]. Graphene's quality clearly reveals itself with a pronounced ambipolar electric field effect; charge carriers can be tuned continuously between electrons and holes where electron mobility remains high even at high concentrations in both electrically and chemically doped devices, which translates to ballistic transport on the sub-micrometre scale. This fact suggests that if graphene is used as a channel material, a transistor allowing extremely high-speed operation and with low electric power consumption could be obtained [73].

Due to graphene's unique properties it has been speculated that a GNS can carry a super-current [73], and it is clear that its theoretical electron transfer rates are superior when contrasted to graphite and CNTs. Furthermore, the fast charge carrier properties of graphene (and other two-dimensional materials) were found not only to be continuous, but to exhibit high crystal quality, in which importantly for graphene charge carriers can travel thousands of inter-atomic distances without scattering [2].

As highlighted above, graphene holds inimitable properties that are superior in comparison to other carbon allotropes of various dimensions and from any other electrode material for that matter, thus 'theoretically' suggesting that graphene is an ideal electrode material that could potentially yield significant benefits in many electrochemical applications; it is this we explore in greater detail in subsequent chapters.

References

1. E. Fitzer, K.-H. Kochling, H.-P. Boehm, H. Marsh, Pure Appl. Chem. **67**, 473–506 (1995)
2. A.K. Geim, K.S. Novoselov, Nat. Mater. **6**, 183–191 (2007)
3. P.R. Wallace, Phys. Rev. **71**, 622–634 (1947)
4. D.A.C. Brownson, D.K. Kampouris, C.E. Banks, Chem. Soc. Rev. **41**, 6944-6976 (2012)
5. K.S. Novoselov, A.K. Geim, S.V. Morozov, D. Jiang, Y. Zhang, S.V. Dubonos, I.V. Grigorieva, A.A. Firsov, Science **306**, 666–669 (2004)
6. K.S. Novoselov, D. Jiang, F. Schedin, T.J. Booth, V.V. Khotkevich, S.V. Morozov, A.K. Geim, Proc. Natl. Acad. Sci. U.S.A. **102**, 10451–10453 (2005)
7. Y. Zhang, Y.-W. Tan, H.L. Stormer, P. Kim, Nature **438**, 201–204 (2005)
8. A.K. Geim, Phys. Scr. **2012**, 014003 (2012)
9. A. Dato, Z. Lee, K.-J. Jeon, R. Erni, V. Radmilovic, T. J. Richardson and M. Frenklach, Chem. Commun. **40**, 6095–6097 (2009)
10. C. Liu, S. Alwarappan, Z. Chen, X. Kong, C.-Z. Li, Biosens. Bioelectron. **25**, 1829–1833 (2010)
11. See summary by, E. S. Reich. Nature **468**, 486 (2010)
12. Web-Resource, The 2010 nobel prize in physics—press release, Nobelprize.org. http://www. nobelprize.org/nobel_prizes/physics/laureates/2010/press.html. Accessed 28 Feb 2012
13. D.R. Dreyer, R.S. Ruoff, C.W. Bielawski, Angew. Chem. Int. Ed. **49**, 9336–9344 (2010)
14. N.R. Gall, E.V. Rutkov, A.Y. Tontegode, Int. J. Mod. Phys. B **11**, 1865–1911 (1997)
15. H.-P. Boehm, A. Clauss, G.O. Fischer, U. Hofmann, Z. Naturforsch. B: Anorg. Chem. Org. Chem. Biochem. Biophys. Biol. **17**, 150–153 (1962)
16. H.-P. Boehm, R. Setton, E. Stumpp, Carbon **24**, 241–245 (1986)

17. H.-P. Boehm, Angew. Chem. Int. Ed. **49**, 9332–9335 (2010)
18. S. Park, R.S. Ruoff, Nat. Nanotechnol. **4**, 217–224 (2009)
19. S. Stankovich, R.D. Piner, X. Chen, N. Wu, S.T. Nguyen, R.S. Ruoff, J. Mater. Chem. **16**, 155–158 (2006)
20. H.-J. Shin, K.K. Kim, A. Benayad, S.M. Yoon, H.K. Park, I.-S. Yung, M.H. Jin, H.-K. Jeong, J.M. Kim, J.-Y. Choi, Y.H. Lee, Adv. Funct. Mater. **19**, 1987–1992 (2009)
21. L.-H. Liu, M. Yan, Nano Lett. **9**, 3375–3378 (2009)
22. C. Berger, Z. M. Song, X.B. Li, X.S. Wu, N. Brown, C. Naud, D. Mayou, T. B. Li, J. Hass, A.N. Marchenkov, E.H. Conrad, P.N. First and W.A. de-Heer, Science **312**, 1191–1196 (2006)
23. X.L. Li, X.R. Wang, L. Zhang, S.W. Lee, H.J. Dai, Science **319**, 1229–1232 (2008)
24. Y. Hernandez, V. Nicolosi, M. Lotya, F.M. Blighe, Z. Sun, S. De, I.T. McGovern, B. Holland, M. Byrne, Y.K. Gun'Ko, J.J. Boland, P. Niraj, G. Duesberg, S. Krishnamurthy, R. Goodhue, J. Hutchison, V. Scardaci, A.C. Ferrari, J.N. Coleman, Nat. Nanotechnol. **3**, 563–568 (2008)
25. S. Stankovich, D.A. Dikin, G.H.B. Dommett, K.M. Kohlhaas, E.J. Zimney, E.A. Stach, R.D. Piner, S.T. Nguyen, R.S. Ruoff, Nature **442**, 282–286 (2006)
26. J.S. Wu, W. Pisula, K. Mullen, Chem. Rev. **107**, 718–747 (2007)
27. A. Dato, V. Radmilovic, Z. Lee, J. Phillips, M. Frenklach, Nano Lett. **8**, 2012–2016 (2008)
28. C. Valles, C. Drummond, H. Saadaoui, C.A. Furtado, M. He, O. Roubeau, L. Ortolani, M. Monthioux, A. Penicaud, J. Am. Chem. Soc. **130**, 15802–15804 (2008)
29. P.W. Sutter, J.I. Flege, E.A. Sutter, Nat. Mater. **406**, 406–411 (2008)
30. D.A.C. Brownson, C.E. Banks, Analyst **135**, 2768–2778 (2010)
31. A.K. Geim, P. Kim, Sci. Am. **298**, 90–97 (2008)
32. N. Mohanty, D. Moore, Z. Xu, T.S. Sreeprasad, A. Nagaraja, A.A. Rodriguez, V. Berry, Nat. Commun. **3**, 844 (2012)
33. D. Yoon, H. Moon, H. Cheong, J.S. Choi, J.A. Choi, B.H. Park, J. Korean Phys. Soc. **55**, 1299–1303 (2009)
34. D. Graf, F. Molitor, K. Ensslin, C. Stampfer, A. Jungen, C. Hierold, L. Wirtz, Nano Lett. **7**, 238–242 (2007)
35. A.C. Ferrari, Solid State Commun. **143**, 47–57 (2007)
36. A.G. Guell, N. Ebejer, M.E. Snowden, J.V. Macpherson, P.R. Unwin, J. Am. Chem. Soc. **134**, 7258–7261 (2012)
37. D.A.C. Brownson, L.C.S. Figueiredo-Filho, X. Ji, M. Gomez-Mingot, J. Iniesta, O. Fatibello-Filho, D.K. Kampouris, C.E. Banks, J. Mater. Chem. A **1**, 5962-5972 (2013)
38. D.A.C. Brownson, S.A. Varey, F. Hussain, S.J. Haigh, C.E. Banks, Nanoscale **6**, 1607-1621 (2014)
39. S. Mao, H. Pu, J. Chen, RSC Adv. **2**, 2643–2662 (2012)
40. D.R. Dreyer, S. Park, C.W. Bielawski, R.S. Rouff, Chem. Soc. Rev. **39**, 228–240 (2010)
41. H.-L. Guo, X.-F. Wang, Q.-Y. Qian, F.-B. Wang, X.-H. Xia, ACS Nano **3**, 2653–2659 (2009)
42. Y. Zhu, S. Murali, W. Cai, X. Li, J.W. Suk, J.R. Potts, R.S. Ruoff, Adv. Mater. **22**, 3906–3924 (2010)
43. M.H. Rümmeli, C.G. Rocha, F. Ortmann, I. Ibrahim, H. Sevincli, F. Börrnert, J. Kunstmann, A. Bachmatiuk, M. Pötschke, M. Shiraishi, M. Meyyappan, B. Büchner, S. Roche, G. Cuniberti, Adv. Mater. **23**, 4471–4490 (2011)
44. C. Soldano, A. Mahmood, E. Dujardin, Carbon **48**, 2127–2150 (2010)
45. D. Chen, L. Tang, J. Li, Chem. Soc. Rev. **39**, 3157–3180 (2010)
46. D.A.C. Brownson, C.E. Banks, Phys. Chem. Chem. Phys. **14**, 8264–8281 (2012)
47. U. Khan, A. O'Neill, M. Lotya, S. De, J.N. Coleman, Small **6**, 864–871 (2010)
48. M. Lotya, P.J. King, U. Khan, S. De, J.N. Coleman, ACS Nano **4**, 3155–3162 (2010)
49. X. Li, W. Cai, L. Colombo, R.S. Ruoff, Nano Lett. **9**, 4268–4272 (2009)
50. X.-M. Chen, G.-H. Wu, Y.-Q. Jiang, Y.-R. Wang, X. Chen, Analyst **136**, 4631–4640 (2011)
51. J.S. Park, A. Reina, R. Saito, J. Kong, G. Dresselhaus, M.S. Dresselhaus, Carbon **47**, 1303–1310 (2009)

52. D.R. Cooper, B. D'Anjou, N. Ghattamaneni, B. Harack, M. Hilke, A. Horth, N. Majlis, M. Massicotte, L. Vandsburger, E. Whiteway, V. Yu, ISRN Condens. Matter Phys. **2012**, 501686 (2012)
53. A.A. Green, M.C. Hersam, Nano Lett. **9**, 4031–4036 (2009)
54. Web-Resource. http://www.nanointegris.com. Accessed 28 Feb 2012
55. W.S. Hummers, R.E. Offeman, J. Am. Chem. Soc. **80**, 1339 (1958)
56. Y. Shao, J. Wang, H. Wu, J. Liu, I.A. Aksay, Y. Lin, Electroanalysis **22**, 1027–1036 (2010)
57. Web-Resource. http://www.graphene-supermarket.com. Accessed 28 Feb 2012
58. A. Reina, X. Jia, J. Ho, D. Nezich, H. Son, V. Bulovic, M.S. Dresselhaus, J. Kong, Nano Lett. **9**, 30–35 (2009)
59. X. Li, Y. Zhu, W. Cai, M. Borysiak, B. Han, D. Chen, R.D. Piner, L. Colombo, R.S. Ruoff, Nano Lett. **9**, 4359–4363 (2009)
60. C. Mattevi, H. Kim, M. Chhowalla, J. Mater. Chem. **21**, 3324–3334 (2011)
61. X. Li, C.W. Magnuson, A. Venugopal, R.M. Tromp, J.B. Hannon, E.M. Vogel, L. Colombo, R.S. Ruoff, J. Am. Chem. Soc. **133**, 2816–2819 (2011)
62. H. Kim, C. Mattevi, M.R. Calvo, J.C. Oberg, L. Artiglia, S. Agnoli, C.F. Hirjibehedin, M. Chhowalla, E. Saiz, ACS Nano **6**, 3614–3623 (2012)
63. K.S. Kim, Y. Zhao, H. Jang, S.Y. Lee, J.M. Kim, K.S. Kim, J.-H. Ahn, P. Kim, J.-Y. Choi, B.H. Hong, Nature **457**, 706–710 (2009)
64. A. Reina, S. Thiele, X. Jia, S. Bhaviripudi, M.S. Dresselhaus, J.A. Schaefer, J. Kong, Nano Res. **2**, 216–509 (2009)
65. Y.-H. Lee, J.-H. Lee, Appl. Phys. Lett. **96**, 083101 (2010)
66. C.P. Jones, K. Jurkschat, A. Crossley, C.E. Banks, J. Iran. Chem. Soc. **5**, 279–285 (2008)
67. L.J. Cote, J. Kim, V.C. Tung, J. Luo, F. Kim, J. Huang, Pure Appl. Chem. **83**, 95–110 (2011)
68. R.L. McCreery, Chem. Rev. **108**, 2646–2687 (2008)
69. M. Pumera, Chem. Rec. **9**, 211–223 (2009)
70. X. Wang, L. Zhi, K. Mullen, Nano Lett. **8**, 323–327 (2008)
71. B. Soodchomshom, Phys. B **405**, 1383–1387 (2010)
72. H.B. Heersche, P. Jarillo-Herrero, J.B. Oostinga, L.M.K. Vendersypen, A.F. Morpurgo, Nature **446**, 56–59 (2007)
73. S. Sato, N. Harada, D. Kondo, M. Ohfuchi, Fujitsu Sci. Tech. J. **46**, 103–110 (2009)
74. K.I. Bolotin, K.J. Sikes, Z. Jiang, M. Klima, G. Fudenberg, J. Hone, P. Kim, H.L. Stormer, Solid State Commun. **146**, 351–355 (2008)

Chapter 2
Interpreting Electrochemistry

This chapter introduces readers to the important aspects of electrochemistry which allow a greater understanding and appreciation of the subject, which can then be applied in later chapters where graphene is utilised as an electrode material.

2.1 Introduction

If you have studied science at undergraduate level you will be familiar with Faraday's law and electrode potentials, and appreciate that electrochemical reactions involve charged species whose energy depends on the potential of the phase that such species are contained within. Consider the following simple reversible redox reaction:

$$O(aq) + ne^-(m) \underset{k_{ox}}{\overset{k_{red}}{\rightleftharpoons}} R^{n-}(aq) \tag{2.1}$$

where O and R are the oxidised and reduced forms of a redox couple in an aqueous media. The electrochemical process as expressed in Eq. (2.1) involves the transfer of charge across the interfacial region of a metallic electrode, termed (m) to indicate the source of electrons, and a solution phase (aq) species. The electrochemical reaction described in Eq. (2.1) only proceeds once a suitable electrode is placed into the solution phase, which acts as a source or sink of electrons. It is important to realise that this reaction involves the transfer of charged species, viz electrons, between the electrode surface (m), and the solution phase species (aq), hence the electrode reaction is an interfacial process. As the electron transfer moves towards equilibrium a net charge separation must develop between the electrode and the solution which creates a potential difference at the solution | electrode interface, which is expressed as ϕ_s and ϕ_m respectively, such that the potential drop across the interface is thus:

$$\Delta\phi = \phi_m - \phi_s \tag{2.2}$$

D. A. C. Brownson and C. E. Banks, *The Handbook of Graphene Electrochemistry*,
DOI: 10.1007/978-1-4471-6428-9_2, © Springer-Verlag London Ltd. 2014

In order to measure such a value a complete conducting circuit is required. However, if another electrode is placed into the solution then you have two electrodes, both monitoring the change of the potential difference at the two electrode I solution interfaces, resulting in meaningless information. The solution is to use one electrode I solution interface and a reference electrode which maintains a fixed potential difference, which leads to the following being realised:

$$E = (\phi_m - \phi_s) + X \qquad (2.3)$$

where E is the potential difference being measured, the term $E = (\phi_m - \phi_s) + X$ refers to the electrode of interest and X is the role of the reference electrode, which is constant. Such measurements are undertaken at equilibrium such that no current is drawn through the cell. The potential E reaches a steady value, E_e which now depends on the relative concentrations of O and R, which can be expressed as:

$$E_e = \Delta\phi_{m/s}(O/R) - \Delta\phi_{m/s}(X) = \Delta\phi_{m/s}(O/R) - 0 \qquad (2.4)$$

where X is a reference electrode such as the Standard Hydrogen Electrode (SHE) or the more commonly utilised Saturated Calomel Electrode (SCE). Conventionally the SHE is defined as exhibiting a potential of zero, allowing one to report the potential of half-cells such as the O/R couple, relative to the SHE. For the process defined in Eq. (2.1), the *Nernst equation* is given by:

$$E_e = E_f^0(O/R) + \left(\frac{RT}{nF}\right)In\left(\frac{[O]}{[R]}\right) \qquad (2.5)$$

for the potential established at the electrode under equilibrium, where E_e is the equilibrium potential of the formal potential E_f^0 and the concentrations of the species O and R at the electrode surface, which, under conditions of equilibrium, are the same as their bulk solution values. Such potential determining equilibrium should be familiar from general undergraduate science classes. Above, R is the universal gas constant (8.314 J K^{-1} mol^{-1}), T is the temperature (in Kelvin), n is the number of electrons transferred in the reaction and F is the Faraday constant ($96,485.33$ C mol^{-1}). Note that in Eq. (2.5) the formal potential is defined as:

$$E_f^0 = E^0(O/R) + \left(\frac{RT}{nF}\right)In\left(\frac{\gamma_O^v}{\gamma_R^v}\right) \qquad (2.6)$$

where E^0 is the standard electrode potential and γ is the relevant activity coefficients. The formal potentials depend on temperature and pressure, as do the standard potentials, but will also have a dependence on electrolyte concentrations, not only of the species involved in the potential determining equilibrium but also on other electrolytes that will be present in the solution since these influence ion activities. The formal potential loses the thermodynamic generality of the standard potentials which are only applicable under specific conditions but enables experimentalists to proceed with meaningful voltammetric measurements.

Equilibrium electrochemistry, viz equilibrium electrochemical measurements, while being of fundamental importance since it allows thermodynamic parameters to be readily obtained (such as reaction free energies, entropies, equilibrium constants and solution pH), it is a rather dry subject and not as exciting as dynamic electrochemistry which is the main thrust of electrochemistry that is used commercially in numerous areas, such as in sensing and energy storage/generation.

If we depart from equilibrium electrochemistry to that of dynamic electrochemistry, consider the following electrochemical process:

$$O(aq) + ne^-(m) \longrightarrow R^{n-}(aq) \tag{2.7}$$

at an electrode which is brought about through the application of a suitability negative potential to the electrode. Note that a second electrode will be needed somewhere in the solution to facilitate the passage of the required electrical current through the solution and a reference electrode will also be required as identified above. The process, as described in Fig. 2.1 occurs in the following general steps. First, the reactant diffuses from the bulk solution to the electrode interface, termed mass transport. Next, a potential is applied into the cell which is different to E_e, the potential difference induces the exchange of electrons between the electrode surface and the species in solution and as such, electrolysis occurs where the magnitude of the current, i, is related to the flux of the species in solution, j, by the following:

$$i = nAFj \tag{2.8}$$

where F is the Faraday constant, n is the number of electrons per molecule involved in the electrochemical process and A is the electrode area. The electron transfer process between the electrode and the O (aq) species takes place via quantum mechanical tunnelling between the electrode and reactant close to the electrode, typically $ca.$ 10–20 Å as the rate of tunnelling falls off strongly with separation since it requires overlap of quantum mechanical wave-functions which describes the electron in the electrode and the electroactive species. Note that the above process is complicated by the reactivity of the electro-active species, the nature (type, geometry) of the electrode surface, the applied voltage and the structure of the interfacial region over which the electron transfer process occurs.

In Eq. (2.8) the units of flux are: moles cm^{-2} s^{-1}, which effectively reflects the quantity of material reaching the electrode surface per second. Akin to homogeneous kinetics, the rate law can be described by:

$$j = k(n)[\text{Reactant}]_0 \tag{2.9}$$

Where $k(n)$ is the nth order rate constant for the electron transfer reaction and $[\text{Reactant}]_0$ is the concentration of the electro-active reactant species at the electrode surface (and not in the bulk solution): when $n = 1$, as commonly encountered, corresponding to a 1st order heterogeneous reaction, the units are cm s^{-1}.

The most common configuration for running dynamic electrochemical experiments involves the use of three electrodes, the working electrode, a counter

Fig. 2.1 Schematic representation of a simple electrode reaction

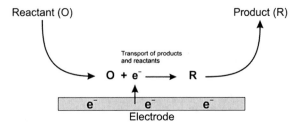

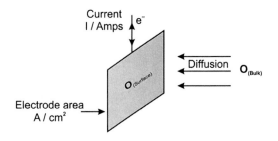

(auxiliary) electrode and a reference electrode, all connected to a commercially available potentiostat which allows the potential difference between the reference and working electrode to be controlled with minimal interference from ohmic (*IR*) drop. The current flowing through the reference electrode is minimised avoiding polarisation of the reference electrode and this keeps the applied potential between the working and reference electrode stable. Shown in Fig. 2.2 is a typical experimental set-up where the three electrode system is being utilised. The reference electrode can be a Ag/AgCl or a SCE which can either be commercially obtained or fabricated within the laboratory. The counter electrode should be a non-reactive high surface area electrode such as platinum or carbon and the working electrode can be a plethora of configurations and compositions, indeed, researchers are trying to utilise graphene as an electrode material, which we will learn about in the subsequent chapters of this *handbook*.

If we concern ourselves with only the working electrode since it is where all the significant processes occur, a general overview of an electrode reaction is depicted in Fig. 2.3 which builds on that shown in Fig. 2.1. This general electrochemical process shows that the observed electrode current is dependent upon mass transport which usually occurs in series with other processes, such as chemical reactions, adsorption/desorption and also the heterogeneous rate constant for the electron transfer reaction. The working electrode is immersed into an electrolyte usually containing the electroactive species under investigation and a supporting electrolyte salt to achieve the required conductivity and to minimise the *IR* drop. The electric double layer at the working electrode occurs over a distance of *ca.* 1 nm. Figure 2.4 shows a schematic representation of the composition of the solution phase close to the (working) electrode surface where the compact layer is also

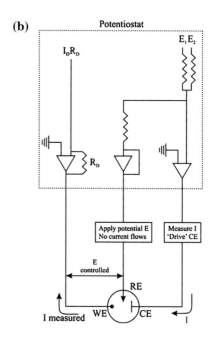

Fig. 2.2 **a** A typical experimental set-up showing the reference electrode (*RE*, saturated calomel electrode), the working electrode (*WE*) and the counter electrode (*CE*, a platinum rod) immersed into an electrolyte solution. **b** A simple electronic scheme equivalent to the electrochemical cell. A potentiostat is required for running electrochemical experiments. Note all the resistances are equal except R_D which is variable

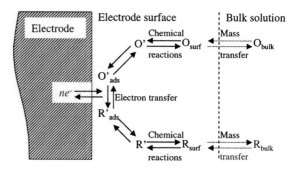

Fig. 2.3 The general electrochemical reaction pathway

termed the "inner Helmholtz" layer which is closest to the surface in which the distribution of charge, and hence potential, changes linearly with the distance from the electrode surface and the diffuse layer, known as the "Gouy-Chapman" layer, in which the potential changes exponentially. Also shown in Fig. 2.4 is a zoomed in perspective of the compact layer showing the Inner and Outer Helmholtz Plane (IHP and OHP) where specifically adsorbed anions and solvated cations can reside.

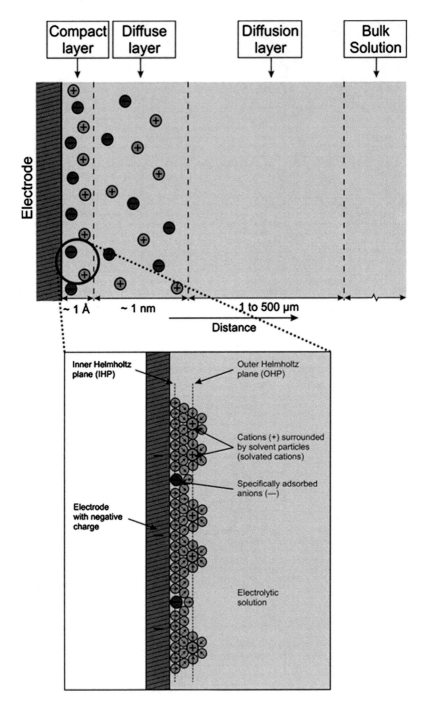

Fig. 2.4 Schematic representation of the composition of the electrode I solution interface (not to *scale*)

Under typical experimental conditions, the magnitude of the diffusion layer is several orders of magnitude larger than the diffuse layer. In dynamic electro-chemistry the potential is always being changed and hence the surface orientation will change and a concentration perturbation travels away from the electrode surface into the solution, where the diffusion layer (δ) is related to the diffusion coefficient of the electroactive species being perturbed as a function of time (t). We return to the diffusion layer when discussing cyclic voltammetry.

As discussed above with reference to Fig. 2.1, the application of a voltage is key for electrochemical reactions to proceed. The application of a potential, that is, a voltage, $V = Joule/Coulombs$, such that the voltage is simply the energy (Joule) required to move charge (Coulomb). The application of such a voltage supplies electrical energy and can be thought of as an electrochemical 'pressure'.

The electronic structure of a metal involves electronic conduction bands in which electrons are free to move throughout the metal, which binds the (metal) cations together. The energy levels in these bands form an effective continuum of levels, which are filled-up to an energy maximum (Fermi level). Such levels can be altered by supplying electrical energy in the form of applying or driving a voltage, as shown in Fig. 2.5a. In Fig. 2.5b (left image), the Fermi level energy is lower than that of the Lowest Unoccupied Molecular Orbital (LUMO) of the reactant and as such it is *thermodynamically unfavourable* for an electron to jump/transfer from the electrode to the molecule. However, as shown in Fig. 2.5b (right image), when the Fermi level of the electrode is above the LUMO of the reactant it is then *thermodynamically favourable* for the electron transfer process to occur, that is, the electrochemical reduction of the reactant can proceed. This concept is explored further in Sect. 2.2 as the process depends on the kinetics of the electrochemical transfer reaction.

2.2 Electrode Kinetics

Let us consider the reduction of iron (III) and the oxidation of iron (II):

$$Fe^{3+}(aq) + e^-(m) \underset{k_{ox}}{\overset{k_{red}}{\rightleftharpoons}} Fe^{2+}(aq) \qquad (2.10)$$

where the rate constants k_{red} and k_{ox} describe the reduction and oxidation respectively. Note that a cathodic process is one at an electrode (a 'cathode') supplying electrons causing a reduction whilst an anodic process is one at an electrode (an 'anode') which removes electrons and causes an oxidation process. The rate law for this net process can be described by the following:

$$j = k_{red}[Fe^{3+}]_0 - k_{ox}[Fe^{2+}]_0 \qquad (2.11)$$

Fig. 2.5 An overview of 'driving' an electrochemical reaction. **a** Band diagram showing the effect of low, medium and high applied voltages. **b** Effect of applied voltage on the Fermi level

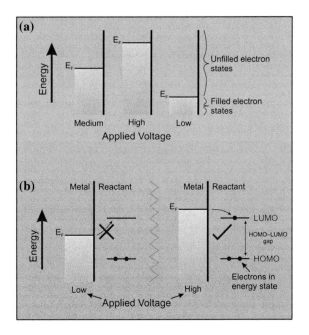

where the rate constants k_{red} and k_{ox} are potential dependent with the cathodic reduction dominating at applied negative electrode potentials whilst the anodic oxidation would be the dominant term at applied positive potentials.

Figure 2.6 depicts a reaction profile for the electrochemical process Eq. (2.10) of interest. The dashed line depicts the energy barrier when no potential has been applied, where it can be seen that the process is thermodynamically uphill. When a potential is applied, the free energy of reactants is raised since the Gibbs free energy for the reduction is related to the formal potential by: $\Delta G^0 = -nF\left(E - E_f^0\right)$ where $\left(E - E_f^0\right)$ measures the potential applied to the working electrode relative to the formal potential of the Fe^{3+}/Fe^{2+} couple with *both* potentials measured relative to the same reference electrode. The reaction coordinate changes to that represented by the solid line where it can be seen that the energy required to reach the transition state is lowered and the process is 'downhill' and is thus thermodynamically driven.

From inspection of Fig. 2.6, we can write:

$$\Delta G_{red}^{\pm}(2) = \Delta G_{red}^{\pm} + \alpha nF\left(E - E_f^0\right) \tag{2.12}$$

and

$$\Delta G_{ox}^{\pm}(2) = \Delta G_{ox}^{\pm} - (1 - \alpha)nF\left(E - E_f^0\right) \tag{2.13}$$

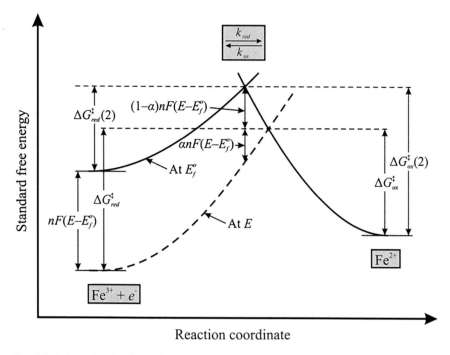

Fig. 2.6 Schematic drawing of the energy profiles along the reaction coordinate for a heterogeneous electron transfer

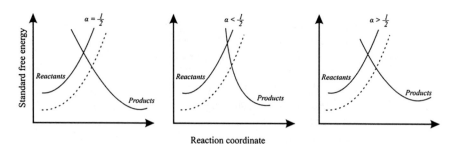

Fig. 2.7 Schematic representation showing the transfer coefficient as an indicator of the symmetry of free energy curve. The dotted line indicates the shift for $Fe^{3+}(aq) + e^-(m) \underset{k_{ox}}{\overset{k_{red}}{\rightleftharpoons}} Fe^{2+}(aq)$ as the potential is altered to more positive potentials

where the parameter α is known as the transfer coefficient, which provides physical insights into how the transition state is influenced by the application of voltage and typically is found to have a value of 0.5. A value of one half means that the transition state behaves mid-way between the reactants and products response to applied voltage. Figure 2.7 shows the effect of changing the potential on the free energy curve. In most systems this lies between 0.7 and 0.3 and usually a value of 0.5 is assumed.

Assuming the rate constants, k_{red} and k_{ox} behave in an Arrhenius form:

$$k_{red} = A_{red} \exp(\Delta G_{red}^{\pm}/RT) \tag{2.14}$$

$$k_{ox} = A_{ox} \exp(\Delta G_{ox}^{\pm}/RT) \tag{2.15}$$

Inserting the activation energies (2.12) and (2.13) gives rise to:

$$k_{red} = A_{red} \exp(\Delta G_{red}^{\pm}(2)/RT) \exp\left(-\alpha n F\left(E - E_f^0\right)/RT\right) \tag{2.16}$$

$$k_{ox} = A_{ox} \exp(\Delta G_{ox}^{\pm}(2)/RT) \exp\left((1 - \alpha)n F\left(E - E_f^0\right)/RT\right) \tag{2.17}$$

since the first part of Eqs. (2.16) and (2.17) are potential independent we can write:

$$k_{red} = k_{red}^0 \exp\left(-\alpha n F\left(E - E_f^0\right)/RT\right) \tag{2.18}$$

$$k_{ox} = k_{ox}^0 \exp\left((1 - \alpha)n F\left(E - E_f^0\right)/RT\right) \tag{2.19}$$

This shows that the electrochemical rate constants for the one electron oxidation of Fe^{2+} (k_{ox}) and for the reduction of Fe^{3+} (k_{red}) depend exponentially on the electrode potential: k_{ox} increases as the electrode is made more positive relative to the solution whilst k_{red} increases as the electrode is made more negative relative to the solution. It is clear that changing the voltage affects the rate constants. However, the kinetics of the electron transfer is not the sole process which can control the electrochemical reaction; in many circumstances it is the rate of mass transport to the electrode which controls the overall reaction, which we diligently explore later.

We know that the net rate (flux) of reaction is given by: $j = k_{red}[Fe^{3+}]_0 - k_{ox}[Fe^{2+}]_0$. Hence using Eqs. (2.18) and (2.19) we can write:

$$j = k_{red}^0 \exp\left[\frac{-\alpha F\left(E - E_f^0\right)}{RT}\right] [Fe^{3+}]_0 - k_{ox}^0 \exp\left[\frac{(1 - \alpha)F\left(E - E_f^0\right)}{RT}\right] [Fe^{2+}]_0 \tag{2.20}$$

If we consider the case of a dynamic equilibrium at the working electrode such that the oxidation and reduction currents exactly balance each other then, since no net current flows, $j = 0$ and the fact that $\alpha = 0.5$, we see that:

$$E = E_f^0 + \frac{RT}{F} \ln\left(\frac{[Fe^{2+}]}{[Fe^{3+}]}\right) + \frac{RT}{F} \ln\left(\frac{k_{ox}^0}{k_{red}^0}\right) \tag{2.21}$$

From the discussion earlier it is clear that when no net current flows the potential is given by:

$$E = E_f^0 + \frac{RT}{F} \ln\left(\frac{[Fe^{2+}]}{[Fe^{3+}]}\right) \tag{2.22}$$

so that $k_{ox}^0 = k_{red}^0 = k^0$ which is the *Nernst equation*. Therefore we can write:

$$k_{red} = k^0 \exp\left[\frac{-\alpha F\left(E - E_f^0\right)}{RT}\right] \tag{2.23}$$

$$k_{ox} = k^0 \exp\left[\frac{(1 - \alpha)F\left(E - E_f^0\right)}{RT}\right] \tag{2.24}$$

Equations (2.23) and (2.24) are the most convenient forms of the *Butler–Volmer expression* for the electrochemical rate constants k_{red}^0 and k_{ox}^0. The quantity k^0, with units of cm s^{-1}, is the *standard electrochemical rate constant*

2.3 Mass Transport

Mass transport of the analyte under investigation is governed by the *Nernst-Planck equation* defined by:

$$J_i(x) = -D_i \frac{\partial C_i(x)}{\partial x} - \frac{z_i F}{RT} D_i C_i \frac{\partial \phi(x)}{\partial x} + C_i V(x) \tag{2.25}$$

where $J_i(x)$ is the flux of the electroactive species i (mol s^{-1} cm^{-2}) at a distance x from the electrode surface, D_i is the diffusion coefficient (cm^2 s^{-1}), $\frac{\partial C_i(x)}{\partial x}$ is the concentration gradient at distance x, $\frac{\partial \phi(x)}{\partial x}$ is the potential gradient, z_i and C_i are the charge (dimensionless) and concentration (mol cm^{-3}) of species i respectively and $V(x)$ is the velocity (cm s^{-1}) with which a volume element in solution moves along the axis. These three key terms comprising Eq. (2.25) represent the contributions to the flux of species i, that is, diffusion, migration and convection respectively.

If we consider an electrochemical experiment which is conducted in a solution that has supporting electrolyte and in stagnant solutions (non-hydrodynamic conditions, see later) such that migration and convection can be neglected from Eq. (2.25), this is thus reduced to consider the only relevant mode of mass transport to the electrode surface on the experimental time scale, which is diffusion.

The diffusion of species i, from bulk solution to the electrode is described by Fick's first and second laws of diffusion;

$$J_i(x) = -D_i \frac{\partial C_i(x)}{\partial x} \qquad (2.26)$$

and

$$\frac{\delta C_i}{\delta t} = -D_i \frac{\partial^2 C_i(x)}{\partial x^2} \qquad (2.27)$$

where j is the flux in mol cm^2 s^{-1}, D the diffusion coefficient in cm^2 s^{-1} and C is the concentration of the electro-active species in mol cm^{-3}. In order to obtain the concentrations of the electro-active species at a location x and time t, the partial differential equation should be solved which is possible if the initial (values at $t = 0$) and boundary conditions (values at certain location x) are known.

Let us consider a simple redox process involving the transfer of one-electron between the electrode and species A in solution to form the product B in solution, as shown below;

$$A + e^- \underset{k_{ox}}{\overset{k_{red}}{\rightleftharpoons}} B \qquad (2.28)$$

where the rate of electron transfer is fast compared to the rate of mass transport, i.e. an electrochemically and chemically reversible redox process. Assuming that the electron transfer follows Butler–Volmer kinetics,

$$k_{red} = k_{red}^0 \exp\left(\frac{-\alpha F}{RT}\eta\right) \qquad (2.29)$$

and

$$k_{ox} = k_{ox}^0 \exp\left(\frac{(1-\alpha)F}{RT}\eta\right) \qquad (2.30)$$

where k^o is the standard electrochemical rate constant, α is the transfer coefficient of the species under investigation and η is the over-potential defined as:

$$\eta = E - E_{A/B}^{0'} \qquad (2.31)$$

where E is the electrode potential and $E_{A/B}^{0'}$ the formal potential for the A/B couple. As electrolysis of A progresses, all of the species A at the electrode surface will be consumed, resulting in a depletion of the concentration of A in the vicinity of the electrode surface and setting up a concentration gradient down which fresh A must diffuse from the bulk solution to support further electrolysis; see Fig. 2.4 which

Fig. 2.8 The Nernst
diffusion layer model

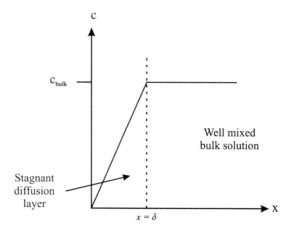

depicts the structure of the electrode surface. This depletion zone is known as the diffusion layer, the thickness of which, δ, increases in size as a function of time, t, such that (in one-dimension):

$$\delta = \sqrt{2Dt} \tag{2.32}$$

Figure 2.8 depicts the Nernst diffusion layer model which shows that beyond the critical distance, δ, the solution is well mixed such that the concentration of the electroactive species is maintained at a constant bulk value. In this vicinity, the mixing of the solution to even out inhomogeneities is due to 'natural convection' induced by density differences. Additionally, if the electrochemical arrangement is not sufficiently thermostated, slight variation throughout the bulk of the solution can provide a driving force for natural convection.

Departing from the bulk solution towards the electrode surface, natural convection dies away due to the rigidity of the electrode surface and frictional forces, this is the diffusion layer, and since only concentration changes occur in this zone, diffusional transport is in operation. Note that in reality there is no real defined zones and these merge into one another, but it is a useful concept. Under experimental conditions, the diffusion layer is in the order of tens to hundreds of micrometers in size.

2.3.1 Cyclic Voltammetry

Cyclic voltammetry is the most extensively used technique for acquiring qualitative information about electrochemical reactions. It tenders the rapid identification of *redox* potentials distinctive to the electroactive species under investigation, providing considerable information about the thermodynamics of a redox process, kinetics of heterogeneous electron-transfer reactions and analysis of coupled electrochemical reactions or adsorption processes. Cyclic voltammetry

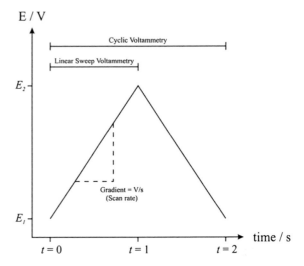

consists of scanning (linearly) the potential of the working electrode using a
triangular potential wave form (Fig. 2.9).

The potential is swept from E_1 to E_2 and the rate at which this is achieved is the
voltammetric scan rate (or the gradient of the line), as shown in Fig. 2.9 (V/s). In
this case, if the potential is stopped, this is known as a linear sweep experiment. If
the scan is returned back to E_1, a full potential cycle, this is known as cyclic
voltammetry. Depending on the information sought, either single or multiple
cycles can be performed. For the duration of the potential sweep, the potentiostat
measures the resulting current that arises via the applied voltage (potential). The
plot of current *versus* potential (voltage) is termed a 'cyclic voltammogram', CV.
A cyclic voltammogram is complex and dependent on time along with many other
physical and chemical properties.

The *cyclic voltammetric* response can be discovered by solving the transport
equations (in three-dimensions, x, y and x) [1]:

$$\frac{\partial[A]}{\partial t} = D_A \nabla^2 [A] \tag{2.33}$$

and

$$\frac{\partial[B]}{\partial t} = D_B \nabla^2 [B] \tag{2.34}$$

and applying Eqs. (2.29) (2.30) and (2.31) as boundary conditions where the
equations [1]:

$$E = E_{start} + vt \quad 0 < t < \frac{E_{end} - E_{start}}{v} \tag{2.35}$$

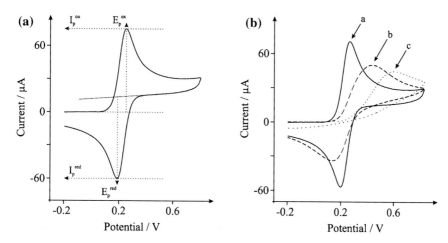

Fig. 2.10 **a** Typical cyclic voltammogram depicting the peak position E_P and peak height I_P. **b** Cyclic voltammograms for reversible (*a*), quasi-reversible (*b*) and irreversible (*c*) electron transfer

and

$$E = E_{end} - \upsilon \left[t - \frac{E_{end} - E_{start}}{\upsilon} \right] \qquad (2.36)$$

define the potential sweep between E_{start} and E_{end} with a voltage sweep rate of, υ, Vs^{-1} and D_A and D_B are the diffusion coefficients of A and B, respectively.

Figure 2.10 shows a typical cyclic voltammetric curve (or CV) for the case of the electrochemical process as described in Eq. (2.28) where a voltammetric potential is applied and the current monitored which gives rise to the unique profile presented in Fig. 2.10a. Characteristics of the voltammogram which are routinely monitored and reported are the peak height (I_P) and the potential at which the peak occurs (E_P).

It is important to note that in plotting voltammetric data, that is current against potential, there are a number of different axis conventions. In the classical (or *polarographic* using mercury) convention, negative potentials are plotted in the positive "x" direction, thus cathodic currents (due to reductions) are positive; this is shown in Fig. 2.11. In the IUPAC convention, the opposite applies where positive potentials are plotted in the positive "x" direction, thus anodic currents (due to oxidations) are positive (in other conventions, current is plotted along the axis and/or a logarithmic current scale is used). In reality the plotting is dictated by the software that is installed on your potentiostat and your geographic location (for example the USA and related countries favour the classical convention). However, one needs to become familiar with this concept as the literature presents voltammograms in mixed styles and one should ensure on first encountering a

Fig. 2.11 The classical convention for plotting voltammetric data

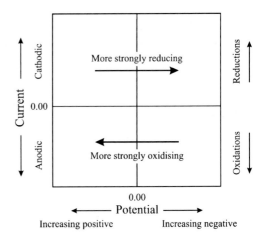

voltammogram that clarification is sought upon which potential sweep has been applied and understand which currents are anodic and which are cathodic.

Shown in Fig. 2.10b is the case of different heterogeneous electron transfer rates, that is, reversible, quasi-reversible and irreversible, each giving rise to unique voltammetric profiles. The physical processes responsible for the characteristic shape of a 'reversible' voltammogram, for the process of $A + ne^- \longrightarrow B$, are based on (i) Fick's laws and (ii) Nernst's laws:

$$\frac{\partial[A]}{\partial t} = D\frac{\partial^2[A]}{\partial x^2} ; \quad \frac{[A]_0}{[B]_0} = e^{\frac{nF\eta}{RT}} \tag{2.37}$$

where the Nernst law is written in the exponential form. It is insightful to consider the diffusion layer at each point in a cyclic voltammetric experiment, which gives rise to the characteristic peak shape observed. Consider the case of electrochemically reversible behaviour. Figure 2.12 shows a typical cyclic voltammogram for $k^o = 1 \ cms^{-1}$ highlighting concentration—distance plots at six different parts on the voltammetric wave.

In the 'reversible' limit the electrode kinetics are so 'fast' (relative to the rate of mass transport—see later) that Nernstian equilibrium is attained at the electrode surface throughout the voltammogram with concentrations of A and B at the electrode surface governed by the Nernst equation:

$$E = E_f^0(A/B) + \frac{RT}{F} In \frac{[B]_0}{[A]_0} \tag{2.38}$$

where E is now the applied potential which defines the ratio of the surface concentrations $[A]_0$ and $[B]_0$ once $E_f^0(A/B)$ is specified.

Figure 2.12 depicts how the concentration profiles and the surface concentrations change during the voltammogram. Point A on the graph corresponds to the

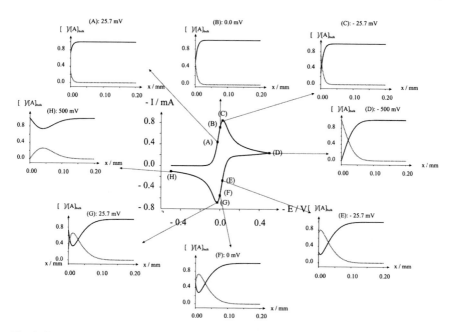

Fig. 2.12 Cyclic voltammogram for the reversible reduction of A to B. Parameters: $E^0 = 0$ V; $\alpha = 0.5$; $k^0 = 1$ cm s^{-1}; $v = 1$ Vs^{-1}; $A = 1$ cm^2; $[A]_0 = 1$ mM; $D_A = D_B = 10-5$ cm^2 s^{-1}. The concentration profiles show the distributions of A (*solid line*) and B (*dashed line*) at eight location, A–H, on the voltammogram. Reproduced with permission from Ref. [10], with permission from Imperial College Press. Note the negative values of the axis (units) labels

formal potential ($E = E_f^0$). At point A, prior to the start of the peak corresponding to the reduction of A, only a small amount of A has been consumed at the electrode surface and only a small layer of B has consequently built up. This diffusion layer is relatively small, typically in the order of *ca.* 10 μm. At point C the maximum reduction current in the voltammetric wave is evident and the diffusion layer has increased in thickness. At point D the current is decreasing with increasing potential and the concentration profile plot shows the concentration of A at the electrode surface to be close to zero so that this part of the voltammogram is under diffusion control whereas at (A) it was the electrode kinetics which controlled the response. The diffusion layer at this point has reached a thickness of *ca.* 40 μm. At point D, the direction of the voltammetric scan is reversed. At point F the working electrode potential has the value of 0 V corresponding to the formal potential of the A/B couple. At this point the electrode potential is insufficient to noticeably reduce A or oxidise B. Point G corresponds to the peak in the reverse scan due to the re-conversion of B to A. The concentration profiles show the build-up of A and depletion of B. Point H corresponds to a point on the reverse peak beyond the maximum G and shows that the concentration of B is very close to zero at the electrode surface whilst that of A has returned to almost its original value nearly that in the bulk solution.

In the case of an electrochemically reversible process *with* fast electron transfer, the peak-to-peak separation $\Delta E_p = \left(E_p^{ox} - E_p^{red}\right)$ is relatively small at the reversible limit, where $\Delta E_p = 2.218RT/nF$, corresponding to a value of *ca.* 57 mV (at 298 K where $n = 1$). For the case of n electrons, the wave-shape of the voltammogram can be characterised by:

$$E_p - E_{1/2} = 2.218\frac{RT}{nF} \tag{2.39}$$

where $E_{1/2}$ corresponds to the potential at which half the peak current is observed.

The magnitude of the voltammetric current $\left(I_p^{Rev}\right)$ observed at a macroelectrode is governed by the following Randles–Ševčik equation:

$$I_p^{Rev} = \pm\, 0.446nFAC(nFD\upsilon/RT)^{1/2} \tag{2.40}$$

where the $\pm$ sign is used to indicate an oxidation or reductive process respectively though the equation is usually devoid of such sign. The voltammetric diagnosis that the electrochemical process is undergoing a reversible heterogeneous charge transfer process is given by Eq. (2.39) where ΔEp is independent of the applied voltammetric scan rate and: $I_p^{ox}/I_p^{red} = 1$.

The question is; *how can you determine if your observed voltammetry corresponds to this range?* A key diagnostic is a scan rate study. As shown in Eq. (2.40), the peak height (I_P) is proportional to the applied voltammetric scan rate and a plot of I_p^{Rev} against $\upsilon^{1/2}$ should be linear. Figure 2.13 depicts typical voltammetric profiles resulting from applying a range of scan rates. It is evident that each voltammetric signature is the same but that the current increases with increasing scan rate as predicted by Eq. (2.40). It is important to note that when the position of the current maximum occurs at the same potential; this peak maximum, which does *not* shift in potential with scan rate, is characteristic of electrode reactions which exhibit rapid electron transfer kinetics, usually termed reversible electron transfer reactions.

The formal potential can be found as the mid-way between the two voltammetric peaks comprising the voltammogram:

$$E_f^0 = \left(E_p^{ox} + E_p^{red}\right)/2 \tag{2.41}$$

assuming that the diffusion coefficients of the reactant and product are equal.

Also shown in Fig. 2.10b is the cyclic voltammetric response for an irreversible electrochemical couple (in which the ΔE_P is larger than that observed for the reversible and quasi-reversible case) where appreciable over-potentials are required to drive the reaction, as evidenced by the peak height (maxima) occurring at a greater potential than that seen for the reversible case.

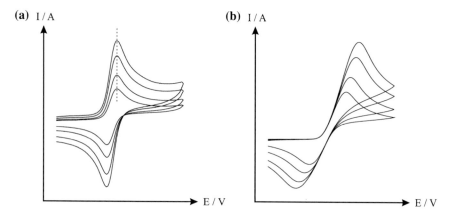

(a) I / A

(b) I / A

E / V

E / V

Fig. 2.13 Reversible (**a**) and irreversible (**b**) cyclic voltammetric responses. Note the shift of the peak maxima with scan rate

In Fig. 2.10 it is evident that as the standard electrochemical rate constant, k^o, is either fast or slow, termed 'electrochemically reversible' or 'electrochemically irreversible' respectively, changes in the observed voltammetry are striking. It is important to note that these are relative terms and that they are in relation to the rate of mass transport to the electrode surface. The mass transport coefficient, m_T, is given by:

$$m_T = \sqrt{D/(RT/Fv)} \tag{2.42}$$

The distinction between fast and slow electrode kinetics relates to the prevailing rate of mass transport given by '$k^o \gg m_T$' indicating electrochemical reversibility or '$k^o \ll m_T$' indicating electrochemical irreversibility. Matsuda and Ayabe [2] introduce the parameter, ζ, given by:

$$\zeta = k^o/(FDv/RT)^{1/2} \tag{2.43}$$

where the following ranges are identified at a stationary macroelectrode: '$\zeta \geq 15$' corresponds to the reversible limit, '$15 > \zeta > 10^{-3}$' corresponds to the quasi-reversible limit and '$\zeta \leq 10^{-3}$' corresponds to the irreversible limit. Thus returning to Fig. 2.10b, we have three cases, reversible, quasi-reversible and irreversible, which are all related to the rate of mass transport. In reversible reactions the electron transfer rate is, at all potentials, greater than the rate of mass transport and the peak potential is independent of the applied voltammetric scan rate (Fig. 2.13a). In the case of quasi-reversible the rate of electron transfer becomes comparable to the mass transport rate. In this regime, the peak potentials increase with the applied scan rate. Last, it is obvious that for the irreversible case the electron transfer rates are smaller than the rate of mass transport; the summary by Matsuda and Ayabe is extremely useful [2].

Fig. 2.14 Transition from a reversible to an irreversible process with increasing scan rate (*solid line*). The *dashed line* indicates a reversible process, while the *dotted line* is that of an irreversible process

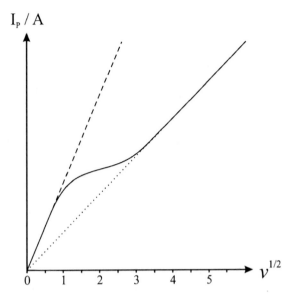

The above conditions given by Matsuda and Ayabe show that the observed electrochemical behaviour depends on the applied voltammetric scan rate. In applying various scan rates the diffusion layer thickness dramatically changes, in the case of slow scan rates, the diffusion layer is very thick while at faster scan rates the diffusion layer is relatively thinner. Since the electrochemical process, that is, reversible or irreversible, reflects the competition between the electrode kinetics and mass transport, faster scan rates will encourage greater electrochemical irreversibility. This is shown in Fig. 2.14 where upon the application of faster scan rates, there is a clear transition (solid line, Fig. 2.14) from that of reversible towards irreversible behaviour.

At macroelectrodes the Nicholson method is routinely used to estimate the observed standard heterogeneous electron transfer rate constant (k^o, cm s^{-1}) for quasi-reversible systems using the following equation [3];

$$\psi = k^o [\pi D n \upsilon F/(RT)]^{-1/2} \qquad (2.44)$$

where ψ is the kinetic parameter and is tabulated (see Table 2.1) at a set temperature for a one-step, one electron process as a function of the peak-to-peak separation (ΔE_P) where one determines the variation of ΔE_P with υ and from this, the variation in the ψ. Table 2.1 shows the variation of ΔE_P with ψ for a one-step, one electron process at 25 °C and where $\alpha = 0.5$. A plot of ψ against $[\pi D n \upsilon F/(RT)]^{-1/2}$ allows the standard heterogeneous rate transfer constant, k^o to be readily deduced.

Note that there are some restrictions, in that the above method is based on the assumption that electron transfer kinetics are described by the Butler–Volmer

Table 2.1 Variation of ΔE_p with ψ at 25 °C. Reproduced with permission from Ref. [3]

ψ	$\Delta E_p \times n$ /mV
20	61
7	63
6	64
5	65
4	66
3	68
2	72
1	84
0.75	92
0.50	105
0.35	121
0.25	141
0.10	212

formalism, that α is 0.5, the switching potential is 141 mV past the reversible $E_{1/2}$, and the temperature is 298 K. Lack of strict adherence to most of these factors will lead to only minor errors. However, there is one experimental problem that can be severe: incomplete compensation of solution resistance. As such, measurement error will be low at slow scan rates where currents and IR errors are low, generally however, potentiostats help overcome this problem.

Beyond the limits of the Nicholson method, that is where the ΔE_P is > 200 mV (see Table 2.1), a suitable relationship has been reported by Klingler and Kochi: [4]

$$k^o = 2.18[D\alpha n\upsilon F/(RT)]^{1/2}\exp\left[-\left(\alpha^2 nF/RT\right)\left(E_p^{ox} - E_p^{red}\right)\right] \qquad (2.45)$$

Thus two procedures are available for different ranges of $\Delta E_P \times n$ values, that is for low (Nicholson) and high values (Klingler and Kochi). Lavagnini et al. [5] proposed the following function of ψ (ΔE_P), which fits Nicholson's data, for practical usage (rather than producing a working curve):

$$\psi = (-0.6288 + 0.021X)/(1 - 0.017X) \qquad (2.46)$$

where $X = \Delta E_P$. For more accurate results in determining k^o, recourse to electrochemical simulation packages is advised.

The Randles–Ševćik equation for a quasi-reversible system (at 298 K) is given by:

$$I_P^{quasi} = \pm\left(2.65 \times 10^5\right)n^{3/2}ACD^{1/2}v^{1/2} \qquad (2.47)$$

For an irreversible system (those with slow electron exchange), the individual peaks are reduced in magnitude and widely separated. Figure 2.13 shows a characteristic response where the peak maximum clearly shifts with the applied

voltammetric scan rate. Totally irreversible systems are quantitatively characterised by a shift in the peak potential with scan rate as given by:

$$E_{p,c} = E_f^0 - \frac{RT}{\alpha n' F}\left[0.780 + In\frac{D^{1/2}}{k^0} + 0.5\,In\left(\frac{\alpha n' Fv}{RT}\right)\right] \qquad (2.48)$$

where α is the transfer coefficient, n' is the number of electrons transferred per mole
before the rate determining step and where E_f^0 is the formal potential. Hence, E_P
occurs at potentials higher than E_f^0, with the over-potential related to k^o and α (the
voltammogram becomes increasingly 'drawn out' as αn decreases). For the case of
a fully irreversible electron transfer process, the Randles–Ševćik equation is:

$$I_p^{irrev} = \pm 0.496(\alpha n')^{1/2} nFAC(FDv/RT)^{1/2} \qquad (2.49)$$

where A is the geometric area of the electrode (cm^2), α is the transfer coefficient
(usually assumed to be close to 0.5), n is the total number of electrons transferred
per molecule in the electrochemical process and n' is the number of electrons
transferred per mole before the rate determining step. It is useful to know the
generic Randles–Ševćik equation (for stagnant solutions):

$$I_p = - \Upsilon(p)\sqrt{\frac{n^3 F^3 v D}{RT}}A[C]$$

$$\text{where : } p = r\sqrt{\frac{nFv}{RTD}} \qquad (2.50)$$

for the case of different electrode geometries:

(1) Planar disc electrode: r = radius, $\Upsilon(p) = 0.446$
(2) Spherical or hemispherical electrode: r = radius, $\Upsilon(p) = 0.446 + 0.752p^{-1}$
(3) For a small disk electrode: r = radius, $\Upsilon(p) = 0.446 + (0.840 + 0.433$
    $e^{-0.66p} - 0.166e^{-11/p})p^{-1} \sim 0.446 + 4/\pi p^{-1}$
(4) For a cylinder or hemi-cylinder: r = radius, $\Upsilon(p) = 0.446 + 0.344p^{-0.852}$
(5) For a band electrode: $2r$ = width, $\Upsilon(p) = 0.446 + 0.614(1+43.6p^2)^{-1} +$
    $1.323p^{0.892} \sim 0.446 + 3.131p^{-0.892}$

The wave-shape for an irreversible reduction is given by: $E_p - E_{1/2} = 1.857\frac{RT}{\alpha F}$,
while for an irreversible oxidation it is given by: $E_p - E_{1/2} = 1.857\frac{RT}{(1-\alpha)F}$.

2.4 Changing the Electrode Geometry: Macro to Micro

At a macroelectrode, electrolysis of A occurs across the entire electrode surface
such that the diffusion of A to the electrode or B from the electrode surface is
termed planar, and the current response is typically described as 'diffusion

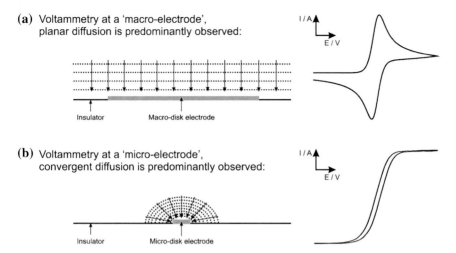

Fig. 2.15 The unique differences between the cyclic voltammetric signatures observed at a macroelectrode (**a**) compared to a microelectrode (**b**)

limited', giving rise to an asymmetric peak as shown in Fig. 2.15a. At the edge of the macroelectrode, where the electrode substrate meets the insulting material defining the electrode area, diffusion to or from the edge of the electrode is effectively to a point. Therefore, the flux, j, and the rate of mass transport are larger at the edge and as such diffusion becomes convergent. This is termed an 'edge effect' which is negligible at a macroelectrode since the contribution of convergent diffusion to the edges of the macroelectrode is inundated by that of planar diffusion to the entire electrode area.

As the electrode size is reduced from macro to micro, or even smaller to that of nano, convergent diffusion to the edges of the electrode becomes significant. In this regime a change in the observed voltammetric profile is observed which results in the loss of the peak shaped response, as evident in Fig. 2.15b with that of a sigmoidal voltammogram. The effect of convergent diffusion has the benefit of improvements in mass transport such that the current density is greater than at a macroelectrode under planar diffusion.

For a reversible electrode reaction at a microelectrode, as shown in Fig. 2.15b, where $E_{1/2}$ is the half-wave potential, the following equation describes the expected voltammetric shape:

$$E = E_{1/2}^{rev} + \frac{RT}{nF} \ln \frac{I_L - I}{I_L}$$

$$\text{where } E_{1/2}^{rev} = E^{0\prime} + \frac{RT}{nF} \ln \frac{D_R^{1/2}}{D_0^{1/2}}$$

(2.51)

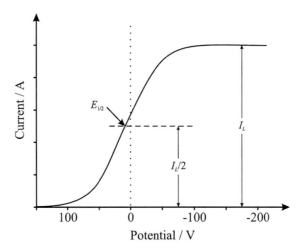

Fig. 2.16 Steady state voltammogram for a reversible process observed at a microelectrode

since the ratio of diffusion coefficient is nearly equal, $E_{1/2}$ is a good approximation for $E^{0\prime}$ for a reversible couple (Fig. 2.16).

When a plot of E against $In\frac{I_L-I}{I_L}$ is constructed, a linear response should be observed with a gradient equal to RT/nF and an intercept of $E_{1/2}^{rev}$ if the wave-shape corresponds to a reversible process. The effect of different electrochemical kinetics is shown for the case of a microelectrode in Fig. 2.17, where the voltammogram is shifted as the electron transfer becomes slower since a greater 'overpotential' is needed to overcome the kinetic barrier. In this case, Eq. (2.51) becomes: $E = E_{1/2}^{irr} + \frac{RT}{\alpha nF} In\frac{I_L-I}{I_L}$. Thus a plot of E against $In\frac{I_L-I}{I_L}$ gives rise to a gradient of $\frac{RT}{\alpha nF}$ and an intercept of $E_{1/2}^{irr}$.

To determine between reversible, quasi-reversible and irreversible, a useful approach is the Tomeš criteria; [7] see Ref. [6] for a full overview of the various diagnostic approaches.

Last, Fig. 2.18 shows the different microelectrode geometries that can be readily encountered in electrochemistry. For an elegant overview of microelectrodes and their benefits and applications, readers are directed to Ref. [7].

2.5 Electrochemical Mechanisms

Above we have considered an E reaction where the (electrochemical) process involves the transfer of an electron. If we now consider that this process is perturbed by a subsequent chemical reaction, as described by:

$$O + ne^- \rightleftharpoons R$$

$$R \rightarrow Z$$

(2.52)

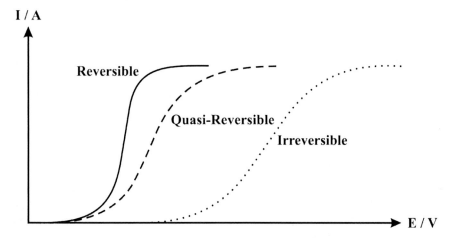

Fig. 2.17 How a steady-state voltammogram is shaped by electrochemical (heterogeneous) kinetics

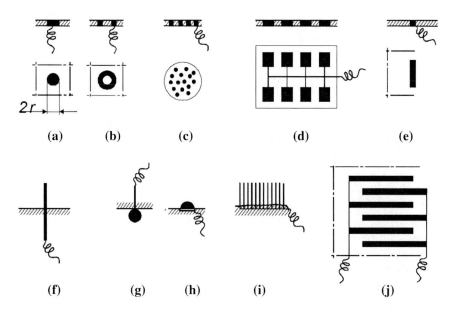

Fig. 2.18 The most important geometries of microelectrodes and microelectrode arrays; **a** microdisk; **b** microring; **c** microdisk array (a composite electrode); **d** lithographically produced microband array; **e** microband; **f** single fibre (microcylinder); **g** microsphere; **h** microhemisphere; **i** fibre array; **j** interdigitated array. Image reproduced with permission from Ref. [7], copyright 2000 International Union of Pure and Applied Chemistry

using the notation of Testa and Reinmuth [8] this is described as an *EC* reaction. The cyclic voltammogram will display a smaller reverse peak (because the product *R* is chemically removed from the surface). The peak ratio of the forward and reverse peaks will thus be less than 1 (not equal); the exact value can be used to estimate the rate constant of the chemical step. In some (extreme) cases, the chemical reaction may progress so rapidly that all of *R* is converted to *Z*, resulting in no reverse wave being observed. Note that by varying the scan rate, further information on the rates of these coupled reactions can be obtained. Table 2.2 overviews the different electrochemical mechanisms involving coupled chemical reactions that can be encountered.

A specific example worth exploring is the *EC'* reaction. Such an example of this process is the modification of a glassy carbon electrode with an Osmium polymer, $[Os(bpy)_2(PVP)_{10}Cl]Cl$ and Nafion prepared by drop-coating, producing a double-layer membrane modified electrode [9]. In this case the modified electrode was explored towards the sensing of the neurotransmitter epinephrine (EP) [9]. Figure 2.19 shows the voltammetric response of the electroactive polymer (curves A and B in Fig. 2.19) where upon contact with epinephrine (curve C in Fig. 2.19) a reduction in the back peak coupled with an increase in the forward wave is evident. The process can be described as:

$$\left[Os - (PVP)_{10}\right]^{+} \leftrightarrows \left[Os - (PVP)_{10}\right]^{2+} + e^{-} \tag{2.53}$$

$$2\left[Os - (PVP)_{10}\right]^{2+} + EP_{RED} \longrightarrow 2\left[Os - (PVP)_{10}\right]^{+} + EP_{OX} \tag{2.54}$$

The first step in the above Eqs. (2.53) and (2.54) is the *E* step due to it being a purely electrochemical process, while the process in Eq. (2.54) is noted as a *C* step due to it being a chemical process. As shown in Fig. 2.19, the magnitude of curve C is dependent upon the chemical rate constant for the process as governed by Eq. (2.54).

Another electrochemical process worth highlighting, which demonstrates how cyclic voltammetry can be used to yield mechanistic information, is an EE process. Here we take the example of TMPD (N,N,N',N'-tetramethylphenylenediamine), the structure of which is shown in Fig. 2.20. Figure 2.21 shows the cyclic voltammogram recorded for the oxidation of TMPD in an aqueous solution (pH 7 phosphate buffer solution, PBS) utilising an EPPG electrode. The two voltammetric peaks, as shown in Fig. 2.21a, represent the following electrochemical process:

$$\begin{aligned} TMPD - e^{-} &\longrightarrow TMPD^{\bullet+} \\ TMPD^{\bullet+} - e &\longrightarrow TMPD^{2+} \end{aligned} \tag{2.55}$$

Table 2.2 Electrochemical mechanisms involving coupled chemical reactions

---

Reversible electron transfer process, no follow-up chemistry; an E_r step:
$$O + ne^- \leftrightarrows R$$
Reversible electron transfer process followed by a reversible chemical reaction; E_rC_r:
$$O + ne^- \leftrightarrows R$$
$$R \underset{k_{1-1}}{\overset{k_1}{\leftrightarrows}} Z$$
Reversible electron transfer profess followed by an irreversible chemical reaction; E_rC_i:
$$O + ne^- \leftrightarrows R$$
$$R \xrightarrow{k_1} Z$$
Reversible chemical reaction preceding a reversible electron transfer process; C_rE_r:
$$Z \underset{k_{1-1}}{\overset{k_1}{\rightleftarrows}} O$$
$$O + ne^- \rightleftarrows R$$
Reversible electron transfer processes followed by an irreversible regeneration of starting materials; $E_rC_i{}'$
$$O + ne^- \leftrightarrows R$$
$$R \xrightarrow{k} Z + O$$
Multiple electron transfer processes with an intervening reversible chemical reaction; $E_rC_rE_r$:
$$O + ne^- \rightleftarrows R$$
$$R \underset{k_{1-1}}{\overset{k_1}{\rightleftarrows}} Z$$
$$Z + ne^- \rightleftarrows Y$$
Multiple electron transfer processes with an intervening irreversible chemical reaction; $E_rC_iE_r$:
$$O + ne^- \rightleftarrows R$$
$$R \xrightarrow{k_1} Z$$
$$Z + ne^- \rightleftarrows Y$$

---

where the cation radical and the dication are shown in Fig. 2.22. On the reverse scan, the corresponding reduction takes place;

$$TMPD^{2+} + e^- \longrightarrow TMPD^{\bullet+}$$
$$TMPD^{\bullet+} + e \longrightarrow TMPD$$
(2.56)

The voltammetric response, as shown in Fig. 2.21b is recorded at a slower scan rate than that used in Fig. 2.21a and it is evident that the second reduction peak, corresponding to the reduction of $TMPD^{2+}$, has significantly changed. This is because in the case of Fig. 2.21b the time taken to scan the voltammetric window is long in comparison with the lifetime of the electro-generated species (formed on the forward scan). In fact, $TMPD^{2+}$ reacts with water with the displacement of

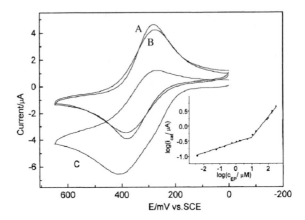

Fig. 2.19 Cyclic voltammograms of Os-(PVP)$_{10}$ (A) and Os-(PVP)$_{10}$/Nafion (B, C) modified electrodes in pH 6.9 PBS (A, B) and (B), + 1.0 × 10^{-4} M epinephrine (C) at a scan rate of 40 mV s^{-1}. Inset: plot of logarithm of catalytic current versus epinephrine concentration. Figure reproduced from Ref. [9] with permission from Elsevier

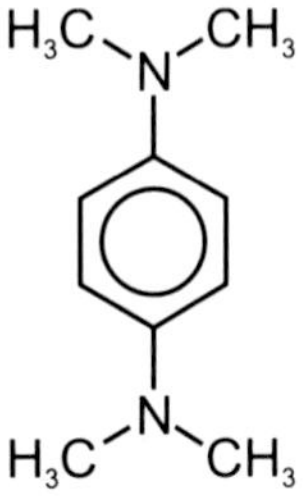

Fig. 2.20 The structure of TMPD

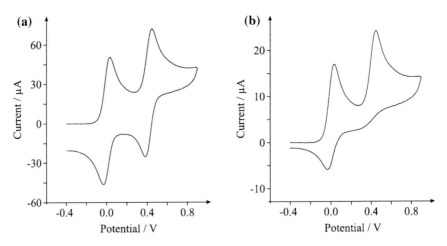

Fig. 2.21 Cyclic voltammograms obtained from the electrochemical oxidation of TMPD at a scan rate of **a** 100 mVs^{-1} and **b** 10 mVs^{-1}

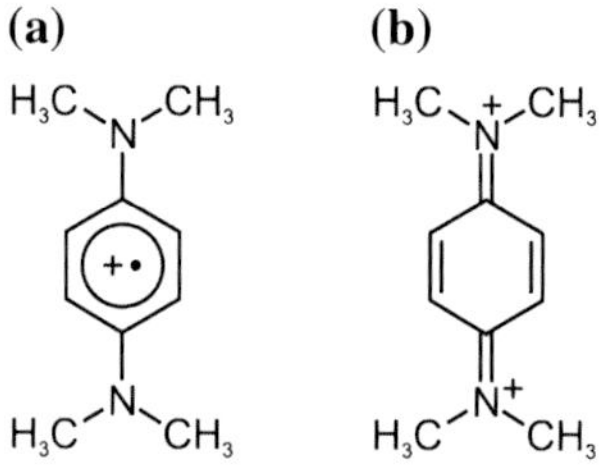

(a)   **(b)**

Fig. 2.22 The structures of **a** the cation radical TMPD^{1+} and **b** the dication TMPD^{2+}

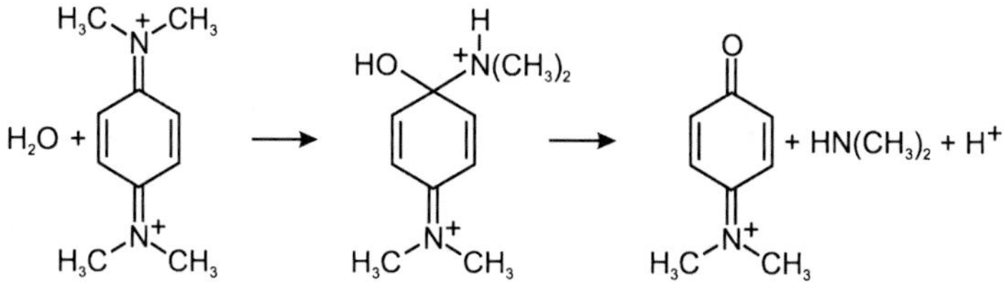

Fig. 2.23 TMPD^{2+} reacts with water with the displacement of dimethylamine

dimethylamine, as shown in Fig. 2.23 and hence on the timescale of the voltammetric experiment the electro-generated species undergoes a chemical reaction such that the initially formed product cannot be electrochemically reduced on the return voltammetric scan. Note that this is not the case when using fast scan rates where the time take to scan the voltammetric window is fast in comparison to the lifetime of the electro-generated species such that the chemical process is outrun.

Hence, given the above insights, it is clear that cyclic voltammetry can be used to provide a facile methodology to study unstable and exotic species.

2.6 Effect of pH

Consider the following process involving the uptake of m-protons and consumption of n-electrons:

$$A + mH^+ + ne^- \leftrightarrows B \tag{2.57}$$

The limiting cases correspond to those of electrochemical reversibility and irreversibility. Here we consider the electrode process being fully electrochemically reversible, thus for the relevant Nernst equation we can write [10]:

$$E = E_f^0(A/B) - \frac{RT}{nF} \ln \frac{[B]}{[A][H^+]^m} \tag{2.58}$$

Fig. 2.24 A typical plot of
peak potential, E_P versus pH

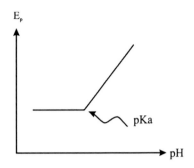

$$E = E_f^0(A/B) + \frac{RT}{nF} In[H^+]^m - \frac{RT}{nF} In\frac{[B]}{[A]} \qquad (2.59)$$

$$E = E_f^0(A/B) - 2.303\frac{mRT}{nF} pH - \frac{RT}{nF} In\frac{[B]}{[A]} \qquad (2.60)$$

leading to:

$$E_{f,\,eff}^0 = E_f^0(A/B) - 2.303\frac{mRT}{nF} pH \qquad (2.61)$$

where $E_{f,\,eff}^0$ is an effective formal potential. Provided $D_A = D_B$ the potential midway between the peaks for the reduction of A and the oxidation of B corresponds to $E_{f,\,eff}^0$ with the shape of the voltammogram being otherwise unaffected. Accordingly the midpoint potential varies by an amount of $2.303\frac{mRT}{nF}$ per pH unit. In the commonly seen case where $m = n$, this corresponds to *ca.* 59 mV per pH unit at 25 °C as in the following case.

Experimentally, the cyclic voltammetric response is recorded over a range of pH's with the $E_{f,\,eff}^0$ (or more commonly 'peak potential') plotted as a function of pH. Figure 2.24 shows a typical response where the deviation from linearity is due to the *pKa* of the target analyte and the gradient from the linear part allows information on the number of electrons and protons transferred in the electrochemical process.

A real example from the literature is shown in Fig. 2.25 which utilises a catechin-immobilised poly(3,4-ethylenedioxythiophene)–modified electrode towards the electrocatalysis of NADH in the presence of ascorbic acid and uric acid [11]. Interestingly, catechin has a quinone moiety in its oxidised state and the effect of pH on the redox properties of the modified electrode is shown in Fig. 2.25 over the pH range of 2–10 where the redox couple of the catechin molecules are shifted to less positive values with the increase in pH. The insert in Fig. 2.25 shows a plot of the half-wave potential of the catechin molecule as a function of pH. Note it

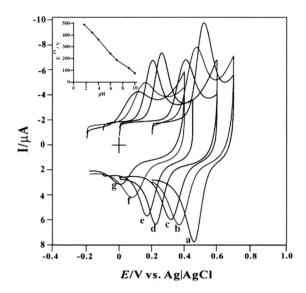

Fig. 2.25 Example from the literature: cyclic voltammograms of the catechin/PEDOT/GC-modified electrode in PBSs of various pH: **a** 1.5; **b** 3; **c** 4; **d** 6; **e** 7; **f** 9 and **g** 10. Inset: plot of $E_{1/2}$ versus pH. Reproduced from Ref. [11] with permission from Elsevier

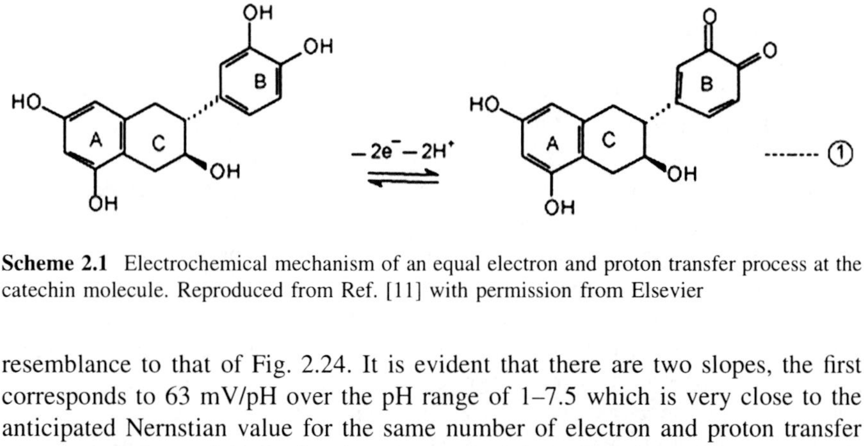

Scheme 2.1 Electrochemical mechanism of an equal electron and proton transfer process at the catechin molecule. Reproduced from Ref. [11] with permission from Elsevier

resemblance to that of Fig. 2.24. It is evident that there are two slopes, the first corresponds to 63 mV/pH over the pH range of 1–7.5 which is very close to the anticipated Nernstian value for the same number of electron and proton transfer processes (Scheme 2.1). In the above example, it is expected that this corresponds to two electrons and two protons involved during the oxidation of catechin to its *o*-quinone form. The second slope is evident in the pH range of 7.5—10; a slope of 33 mV/pH is obtained which is very close to the Nernstian value for a two-electron and one-proton process (Scheme 2.2). Additionally, the authors [11] report a gradual decreases in the peak currents in this pH range which they attribute to the deprotonation of catechin molecules, which leads to an increase in the charge on the catechin molecules where the charged species is more soluble than the former

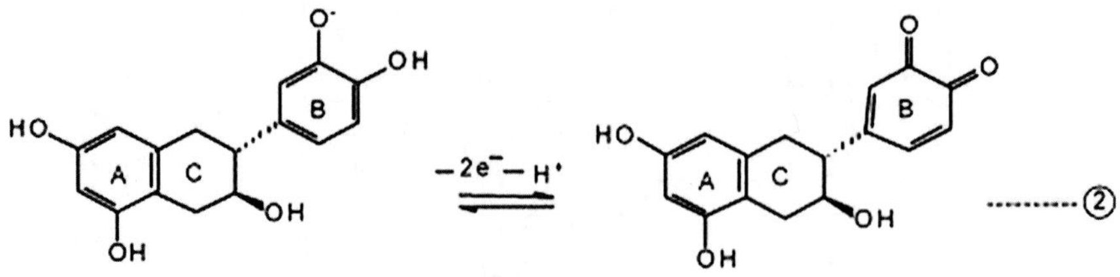

Scheme 2.2 Electrochemical mechanism of a two-electron and one-proton transfer process at the catechin molecule. Reproduced from Ref. [11] with permission from Elsevier

species. Clearly the use of pH measurements can help provide insights into electrochemical mechanisms.

2.7 Other Voltammetric Techniques: Chronoamperometry

In the above part of the *Handbook,* we have considered cyclic voltammetry and its derivatives. Another technique that is worthy of mention, and that can be used to study graphene, is chronoamperometry. This technique is commonly used either as a *single potential step*, in which only the current resulting from the forward step (as described above) is recorded, or *double potential step*, in which the potential is returned to a final value following a time period, usually designated as τ, at the step potential. The electrochemical technique of chronoamperometry involves stepping the potential applied to the working electrode, where initially it is held at a value at which no Faradaic reactions occur before jumping to a potential at which the surface concentration of the electroactive species is zero (Fig. 2.26a), where the resulting current-time dependence is recorded (Fig. 2.26c).

The mass transport process throughout this process is solely governed by diffusion, and as such the current-time curve reflects the change in concentration at the electrodes surface. This involves the continuing growth of the diffusion layer associated with the depletion of reactant, thus a decrease in the concentration gradient is observed as time progresses (Fig. 2.26b). An example of single potential step chronoamperometry is shown in Fig. 2.27 for the case of an Osmium complex modified electrode and also shown in the inserts are the effect of concentration of epinephrine which is electro-catalysed undergoing an EC' process (see Sect. 2.5).

The most useful equation in chronoamperometry is the Cottrell equation, which describes the observed current (planar electrode of infinite size) at any time following a large forward potential step in a reversible redox reaction (or to large overpotential) as a function of $t^{-1/2}$.

$$I_L(t) = nFAD^{1/2}C(\pi t)^{-1/2} \tag{2.62}$$

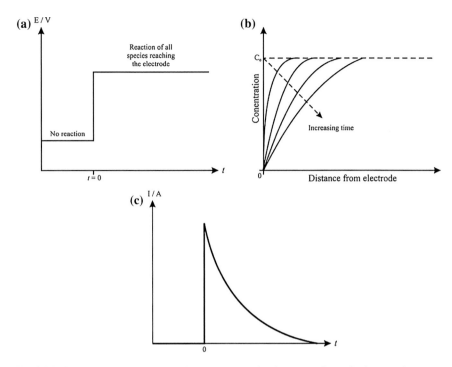

Fig. 2.26 Chronoamperometric experiment: **a** potential-time waveform; **b** change of concentration gradient; **c** resulting current-time response

where n = stoichiometric number of electrons involved in the reaction, F is the Faraday's constant, A is the electrode area, C is the concentration of electroactive species and D is the diffusion coefficient. The current due to double layer charging also contributes following a potential step but decays as a function of $1/t$ and is only significant during the initial period, typically a few milliseconds following the potential step.

A variant on the chronoamperometry as discussed in Fig. 2.27 is that presented in Fig. 2.28, which is where a hybrid biointerface electrode, consisting of a gold nanoparticle (AuNP) and *cytochrome c* (*cyt c*) on indium tin oxide (ITO) platform, is explored towards the sensing of hydrogen peroxide. In this example, the potential is held at a value, which induces the desired electrochemical reaction, and the solution is stirred such that a convective flow is induced. Aliquots of the analyte under investigation are made and at each time, the convective flow ensures that the species is transported to the electrode surface and is electrochemically transformed; this is recognised in Fig. 2.28 by a 'step' in the current which then plateaus or reduces as all the electroactive species are consumed. Analysis of the 'steps', as evident in Fig. 2.28, yields the corresponding calibration plot. Such an approach, where convection is used to enhance the rate of mass transport to the electrode surface, is known as hydrodynamic electrochemistry and offers

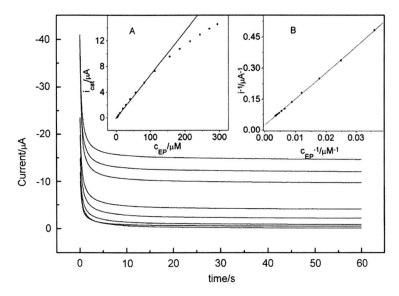

Fig. 2.27 Example from the literature: chronoamperometric curves with a potential step from 0 to +0.4 V at Os-(PVP)$_{10}$/Nafion modified electrode in pH 6.9 PBS containing 0, 6.5, 11, 28, 56, 156, 215 and 294 µM (from *bottom* to *top*). Inset: *A* plot of catalytic current versus epinephrine concentration; *B* data analysis of catalytic current versus epinephrine concentration. Reproduced from Ref. [9] with permission from Elsevier

improvements in the analytical sensitivity in comparison to measurements performed in stagnant solution (Fig. 2.27).

Double potential step chronoamperometry is illustrated in Fig. 2.29 for the electrochemical oxidation of ferrocene (Fc) to ferrocenium (Fc$^+$). Double potential step chronoamperometry is extremely useful as it allows simultaneous determination of the diffusion coefficients of the initial species, in this example, ferrocene, and of the product of the electrochemical reaction, ferrocenium. Figure 2.29 shows typical chronoamperometry currents recorded at different temperatures using a microelectrode where the sample was pre-treated by holding the potential at a point corresponding to zero Faradaic current for 20 s, after which the potential was stepped to a position after the peak (as determined via CV) and the current measured for 5 s. The potential was then stepped back to the initial value and the current measured for a further 5 s. Note that for chronoamperometry at a microelectrode, Eq. (2.62) is replaced by:

$$I_L(t) = nFADC\left[(\pi Dt)^{-1/2} + r^{-1}\right] \tag{2.63}$$

which turns back into Eq. (2.62) if $r \longrightarrow \infty$.

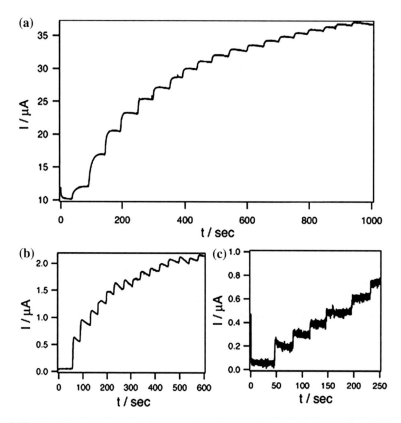

Fig. 2.28 Example from the literature: current–time curve obtained for **a** ITO/AuNP/*cyt c*
electrode upon successive addition of 20 μL aliquots of 200 mM H$_2$O$_2$ to 5 mL stirred 10 mM
HEPES buffer at pH 7 with an applied potential of −0.1 V under nitrogen atmosphere;
chronoamperometric curve obtained for **b***cyt c*/ITO and **c** AuNP/ITO obtained by the addition of
20 μL aliquots of 200 mM H$_2$O$_2$ in 5 mL stirred solution of 10 mM HEPES buffer at the potential
of −0.1 V under nitrogen atmosphere. Reproduced from Ref. [29] with permission from Elsevier

The time-dependent current response obtained on the first step (see Fig. 2.29)
was analysed via the use of the following equation, as proposed by Shoup and
Szabo [12], which sufficiently describes the current response over the entire time
domain, with a maximum error of less than 0.6 %:

$$I = -4nFDrCf(\tau) \tag{2.64}$$

where

$$f(\tau) = 0.7854 + 0.8863\tau^{-1/2} + 0.2146\exp(-0.7823\tau^{-1/2}) \tag{2.65}$$

and where the dimensionless time parameter, τ is given by:

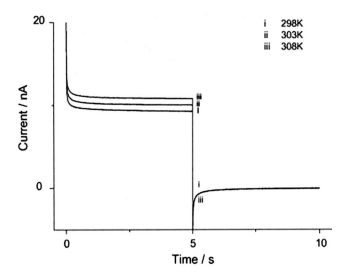

Fig. 2.29 Experimental double potential step chronoamperometric transients for the Fc|Fc$^+$ system at temperatures of *i* 298 *ii* 303 and *iii* 308 K. The potential was stepped from 0.0 to 0.9 and back to 0.0 V. Reproduced from Ref. [30] with permission from Elsevier

$$\tau = 4Dt/r^2 \tag{2.66}$$

Theoretical transients were generated using the above equation and a nonlinear curve fitting function available in certain software (such as OriginPro 7.5—Microcal Software Inc.). The fit between the experimentally observed response and theoretical data was optimised by inputting a value for r, which is independently characterised using model solutions and instructing the software to iterate through various D and nc values. Figure 2.30 shows a typical fit of experimental and simulated results using Eqs. (2.64), (2.65) and (2.66).

2.7.1 Experimental Determination of Diffusion Coefficients

The double potential step method is often applied in the measurement of rate constants for chemical reactions (including product adsorption) occurring following the forward potential step. Chronoamperometry can also be used to determine an accurate measurement of electrode area (A) by use of a well-defined redox couple (known n, C, and D). Other uses are in extracting nucleation rates (see for example: [13–17]) and uses in sensing, as shown in Figs. 2.27 and 2.28, to name just a few. In some instances, when one is experimentally measuring a diffusion coefficient of a target analyte which has not been reported before, one needs to know that the

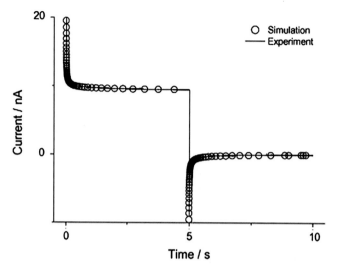

Fig. 2.30 Experimental (–) and best-fit theoretical (○) double potential step chronoampero-metric transients for the Fc|Fc$^+$ system at a temperature of 298 K in MeCN with 0.1 M TBAPF$_6$ supporting electrolyte. The potential was stepped from 0.0 to 0.9 and back to 0.0 V. Reproduced from Ref. [30] with permission from Elsevier

correct number (or order of magnitude) has been derived. To this end, a model for estimating the diffusion coefficients has been proposed by Wilke and Chang: [18]

$$D = 7.4x10^{-8} \frac{(xM)^{1/2}T}{\eta V^{0.6}} \tag{2.67}$$

where M is the relative molar mass of the solvent, η is its viscosity, x is an association parameter and V is the molar volume of the solute, estimated for complex molecules by summation of atomic contributions [18].

2.8 Other Voltammetric Techniques: Differential Pulse Voltammetry

As introduced earlier, voltammetry so far has been concerned with applying a potential step where the response is a pulse of current which decays with time as the electroactive species near the vicinity of the electrode surface are consumed. This Faradaic process (I_F) is superimposed with a capacitative contribution (I_C) due to double layer charging which dies away much more quickly, typically within microseconds (see Fig. 2.31).

The current (for a reversible system) is in the form of the Cottrell Equation where $I \propto t^{-1/2}$ and charge, Q, is $Q \propto t^{-1/2}$. When a step in pulse is applied the

Fig. 2.31 Effect of capacitative and Faradaic current following the application of a potential step

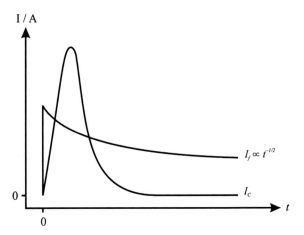

current is sampled when the capacitative current (I_C) decays away. To achieve this pulse widths are chosen to meet this condition.

In pulse techniques such as differential pulse and square wave voltammetry, the capacitative contribution is eliminated via subtraction. Differential pulse voltammetry (DPV) measures the difference between two currents just before the end of the pulse and just before its application. Figure 2.32 shows the waveform of pulse utilised which is superimposed on a staircase.

The base potential is implemented in a staircase and the pulse is a factor of 10 or more shorter than the pulse of the staircase waveform. The difference between the two sampled currents is plotted against the staircase potential leading to a peak shaped waveform as shown in Fig. 2.33.

For a reversible system the peak occurs at a potential: $E_p = E_{1/2} - \Delta E/2$ where ΔE is the pulse amplitude. The current is given by:

$$I_p = \frac{nFAD^{1/2}C}{\pi^{1/2}t^{1/2}} \left(\frac{1-\alpha}{1+\alpha}\right) \tag{2.68}$$

where

$$\alpha = \exp(nF\Delta E/2RT) \tag{2.69}$$

DPV is useful due to eliminations in the contribution of non-faradaic (capacitive) processes, which are effectively subtracted out. The power of DPV is evident from inspection of Fig. 2.34 for the supercoiled plasmid DNA where poor electrochemical signals obtained using linear sweep voltammetry are transformed into quantifiable and beautiful voltammetric signatures.

Additionally, DPV is useful for resolving the voltammetric signals due to two species with close half-wave potentials, producing easily quantifiable peak shaped responses. This is exemplified in Fig. 2.35 for the simultaneous sensing of ascorbic

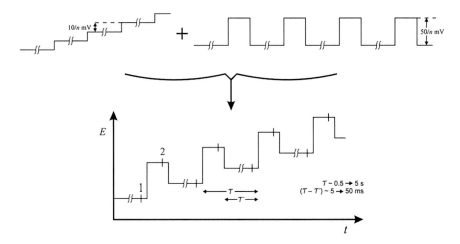

Fig. 2.32 Differential pulse voltammetry waveform of pulses superimposed on a staircase

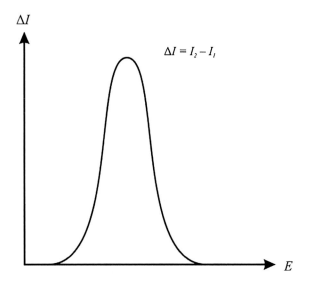

Fig. 2.33 Differential pulse voltammetry; voltammetric profiles of ΔI versus staircase potential

acid and acetaminophen which are well known to cause problems due to their overlapping voltammetric responses.

Figure 2.35 compares CV and DPV using a range of modified electrodes. In all cases, two sharp and well-resolved peaks are observed when DPV is utilised; in this analytical case, the resulted separation in the two peak potentials is sufficient enough to achieve the accurate simultaneous determination of ascorbic acid and acetaminophen in real samples [19].

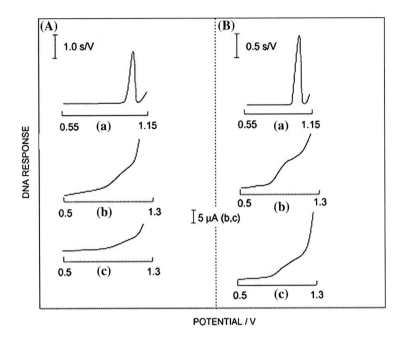

Fig. 2.34 Differential pulse (*a*) and linear scan (*b*, *c*) voltammograms for 15 μg mL^{-1} supercoiled plasmid DNA (**A**) and 15 μg mL^{-1} linearised DNA (**B**) at carbon paste electrodes. The anodic signal corresponds to the electrochemical oxidation of DNA–G residues. Reproduced from Ref. [31] with permission of Elsevier

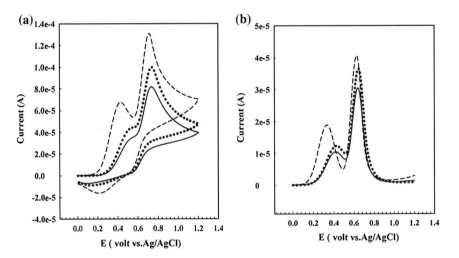

Fig. 2.35 a Cyclic and **b** differential pulse voltammograms of 0.1 mM ascorbic acid and 0.1 mM acetaminophen in acetate buffer solution (0.1 M, pH 4.0) on the surface of various electrodes; unmodified carbon paste electrode (*solid line*), CNT–carbon paste electrode (*dotted line*) and multi-walled carbon nanotube/thionine modified electrode (*dashed line*). Sweep rate was 100 mV s^{-1}. Reproduced from Ref. [19] with permission of Elsevier

Note that such a response is obtained through increasing the pulse amplitude (see Fig. 2.32) but in doing so, the peak width also increases, meaning that, in practice, ΔE values of more than 100 mV are not viable; careful optimisation of the electrochemical parameters is clearly required. The expression for the half-width at half-height, $W_{1/2}$; where $W_{1/2} = 3.52RT/nF$ leads to a value of 90.4 mV for $n = 1$ at 298 K showing that peaks separated by 50 mV may often be resolved. Detection limits of DPV can be realised at *ca.* 10^{-7} M.

Recently it has been highlighted that the literature is not consistent and that actually the waveform presented in Fig. 2.32, should be termed "Differential Multi Pulse Voltammetry" [20]. Figure 2.36 shows the range of waveforms that can exist.

Differential Double Pulse Voltammetry (DDPV), where the length of the second pulse (t_2) is much shorter than the length of the first pulse (t_1), $t_1/t_2 = 50$–100 (Fig. 2.36a), which leads to very high sensitivity. *Differential Double Normal Pulse Voltammetry (DDNPV)* is where both pulses have similar durations $t_1 \approx t_2$ (Fig. 2.36b).

Double Pulse Square Wave Voltammetry (DPSWV) is where both pulses are equal $t_1 = t_2$ and the pulse height ($\Delta E = E_2 - E_1$) is opposite from the scan direction (Fig. 2.36c). Because of its analogy with the potential-time program applied in Square Wave Voltammetry, it is referred to as Double Pulse Square Wave Voltammetry. *Differential Multi Pulse Voltammetry (DMPV)* is as a variant of DDPV where the initial conditions are not recovered during the experiment (Fig. 2.36d). Thus, the pulse length (t_p) is much shorter than the period between pulses (t_1), $t_1/t_p = 50$–100.

Differential Normal Multi Pulse Voltammetry (DNMPV) is the multi-pulse variant of the DMPV technique such that the duration of the period between pulses and the duration of the pulses are similar: $t_1 \approx t_p$ (Fig. 2.36e). Last, *Square Wave Voltammetry* can be considered as a particular situation of DNMPV where the length of both pulses are equal ($t_1 = t_p$) and the sign of the pulse height (ΔE) is opposite from the scan direction (Fig. 2.36f).

2.9 Other Voltammetric Techniques: Square Wave Voltammetry

The square wave voltammetric waveform consists of a square wave superimposed on a staircase, as shown in Fig. 2.37. The currents at the end of the forward and reverse pulses are both registered as a function of staircase potential. The difference between them, the net current, is larger than either of its two component parts in the region of the peak which is centred on the half-wave potential. Capacitive contributions can be effectively discriminated against before they die away, since, over a small potential range between forward and reverse pulses, the capacitance is constant and is thus annulled by subtraction. In this way the pulses can be shorter than in DPV and the square wave frequency can be higher. Instead of the effective sweep rates of 1–10 mVs^{-1} of DPV, scan rates of 1 Vs^{-1} can be employed.

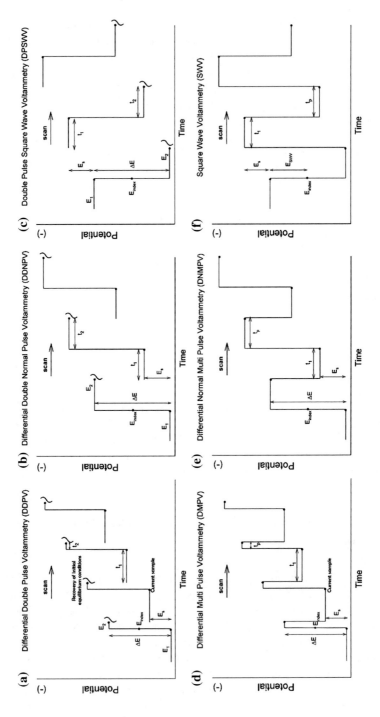

Fig. 2.36 Potential-time programs of the differential pulse techniques considered. Reproduced from Ref. [20] with permission from Elsevier

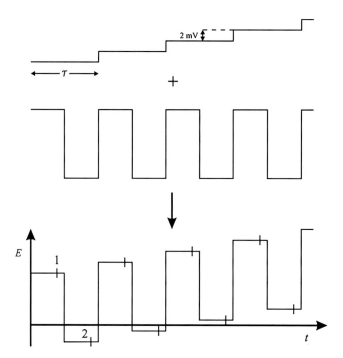

Fig. 2.37 Square wave voltammetry: waveform showing the summation of a staircase and a square wave

Detection limits of *ca.* 10^{-8} M or lower are readily achievable under optimum conditions. The advantages over cyclic voltammetry are as follows: faster scan rates are possible (faster reactions can be studied), higher sensitivity (lower concentrations can be used) and a higher dynamic range (a larger range of concentrations can be investigated). Usually in electrochemistry, solutions are vigorously degassed with, for example, nitrogen to remove oxygen which is electrochemically reduced and can interfere with the voltammetric measurement under investigation. A different way of greatly diminishing or eliminating the interference of oxygen, with no need for its removal, is by the use of the high frequencies employed in SWV. In fact, due to the irreversibility of oxygen reduction, the increase of its signal with frequency is small at high frequencies, and becomes negligible eventually, when compared with the response of the determinant [21] (Fig. 2.38).

2.10 Other Voltammetric Techniques: Stripping Voltammetry

Anodic Stripping Voltammetry (ASV) is an extremely sensitive electro-analytical technique that can determine trace quantities of certain metals at the parts-per-billion level. The first phase of an ASV experiment involves a pre-concentration

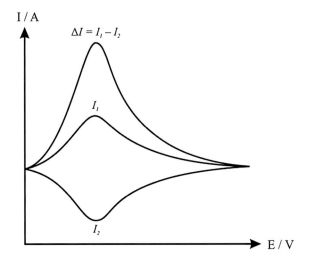

Fig. 2.38 Square wave voltammetry: voltammetric profile of current versus staircase potential. I_1 represents the forward and I_2 the reverse sweep where ΔI is the resultant voltammogram

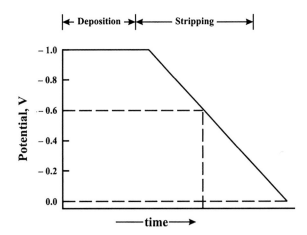

Fig. 2.39 A potential *vs.* time profile showing the stages in ASV

step, in which the analyte is deposited (i.e. in the case of metal analysis, reduced to its elemental form) at the working electrode by controlled potential electrolysis in a stirred solution at a suitable reduction potential as shown in Fig. 2.39. The process can be written as:

$$M^{n+}(aq) + ne^- \rightarrow M(electrode) \tag{2.70}$$

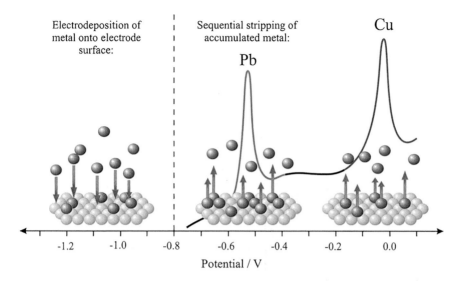

Fig. 2.40 A schematic representation of anodic stripping voltammetry showing the two key steps: electrodeposition and stripping

In the second phase, the potential of the working electrode is scanned so that the deposited metal is oxidised back to its ionic form, i.e. is anodically stripped from the electrode (see Fig. 2.39) by scanning the potential to a value where the following process occurs:

$$M(electrode) \rightarrow M^{n+}(aq) + ne^{-} \tag{2.71}$$

Figure 2.40 shows a schematic representation of this process for the case of lead and copper.

Additionally, as shown in Fig. 2.41, the potential—time profile can be used in ASV with a cleaning step (step A). This is usually applied in-between measurements to ensure that the deposited metal is fully stripped from the electrode surface so as to improve the reproducibility of the analytical measurement.

Figure 2.42 shows the response of a range of metals which occur at different stripping potentials allowing a multitude of metals to be readily analysed at once. The peak area and/or peak height of each stripping signal is proportional to concentration, allowing the voltammetric signal to be used analytically.

In addition to anodic stripping voltammetry there are also cathodic and adsorptive stripping voltammetry. In each case, similar to anodic stripping voltammetry, there is first a pre-conditioning step, which for cathodic stripping is either:

$$2M^{n+}(aq) + mH_2O \rightarrow M_2O_m(electrode) + 2mH + 2(m-n)e^{-} \tag{2.72}$$

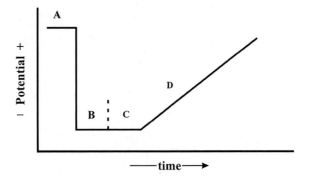

Fig. 2.41 A typical experimental potential—time profile used in ASV. Step *A* is the 'Cleaning step', *B* 'electrodeposition, *C* 'Equilibration step', *D* 'Stripping step'

Fig. 2.42 Stripping voltammetry of zinc, cadmium and lead all in the same aqueous solution

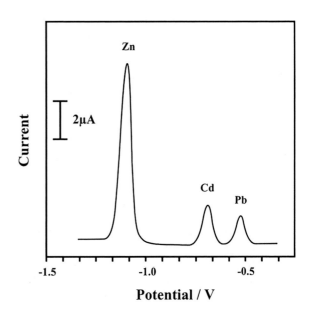

or

$$M + L^{m+} \rightarrow ML^{(n-m)-}(ads) + ne^- \qquad (2.73)$$

while for adsorptive stripping:

$$A(aq) \rightarrow A(ads) \qquad (2.74)$$

The corresponding stripping steps are for cathodic either:

$$M_2O_m(electrode) + 2mH + 2(m-n)e^- \rightarrow 2M^{n+}(aq) + mH_2O \qquad (2.75)$$

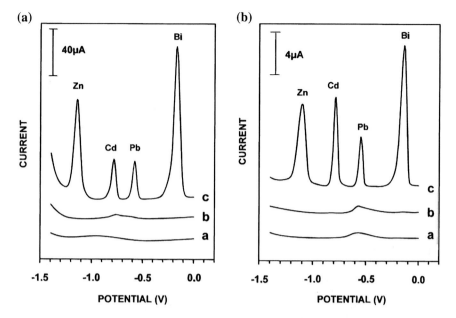

Fig. 2.43 Stripping voltammograms at glassy-carbon (**a**) and carbon-fiber (**b**) electrodes: *a* 0.1 M acetate buffer (pH 4.5); *b* as in (*a*) but after the addition of 50 μg/L Cd(II), Pb(II), and Zn(II); *c* as in (*b*) but after the addition of 400 μg/L Bi(III). Deposition for 120 s at −1.4 V; "cleaning" for 30 s at +0.3 V. Square-wave voltammetric stripping scan with a frequency of 20 Hz, potential step of 5 mV, and amplitude of 25 mV. Reproduced with permission from Ref. [32], copyright 2000 The American Chemical Society

or

$$ML^{(n-m)-}(ads) + ne^- \rightarrow M + L^m \tag{2.76}$$

while for adsorptive stripping:

$$A(ads) \pm ne^- \rightarrow B(aq) \tag{2.77}$$

To impart improvements in the voltammetric response, different electrode materials (and sizes to induce improvements in mass transport) can be used and traditionally mercury film and drop electrodes were utilised. Additionally, the use DPV and SWV can also provide benefits to reduce analytical limits of detection and improve sensitivities. However, due to the toxicity of mercury, considerable efforts have been devoted to the investigation of alternate electrode materials. Non-mercury electrodes, such as gold, carbon or iridium, have been explored but the overall performance of these alternatives has not approached that of mercury [22]. A verified alternative is the use of bismuth film electrodes (on various carbon substrates) which offer high-quality stripping performance that compares favourably with that of mercury electrodes [22, 34].

Shown in Fig. 2.43 is the stripping performance of an in-situ modified bismuth-film electrode which also displays the corresponding control experiments (bare/unmodified electrode). No stripping signals are observed at the bare glassy-carbon (A(b)) and carbon-fibre (B(b)) electrodes for a sample containing 50 μg/L lead and cadmium. In contrast, adding 400 μg/L bismuth to the sample, and simultaneously depositing it along with the target metals, resulted in the appearance of sharp and undistorted stripping peaks for both analytes, as well as for bismuth (c).

The use of bismuth modified electrodes (through ex-situ and in-situ) modification has become a backbone of the electroanalytical community for a variety of target analytes [23].

2.11 Adsorption

In some instances, rather than having the analyte under investigation undergoing simply diffusional processes, the species of interest might adsorb onto the electrode surface and will give rise to different voltammetry. Figure 2.44 shows a typically votlammetric profile where a unique shape is observed. Since the adsorbed species does not have to diffuse to the electrode surface, the observed voltammogram is symmetrical.

The peak current can be related directly to the surface coverage (Γ) and potential scan rate for a reversible process:

$$I_p = \frac{n^2 F^2 \Gamma A \upsilon}{4RT} \tag{2.78}$$

with integration of the peak(s) shown in Fig. 2.44, allows the charge (Q) to be deduced, which is related to the surface coverage by the following expression:

$$Q = nFA\Gamma \tag{2.79}$$

As shown in Fig. 2.44, the full width at half of the peak maximum height (FWHM) is given by:

$$FWHM = 3.53RT/nF \tag{2.80}$$

The diagnosis of an adsorbed species is to explore the effect of scan rate on the voltammetric response, which should yield a linear response for the case of I_p versus scan rate υ. A practical example is shown in Fig. 2.45 where Hemoglobin (Hb)—Dimyristoyl phosphatidylcholine (DMPC) films are immobilised upon a BPPG surface. The modified electrode was then explored with the voltammetric response evident in Fig. 2.45a, with a plot of peak current against scan rate also

Fig. 2.44 Cyclic voltammetric response for the reversible reaction of an adsorbed species

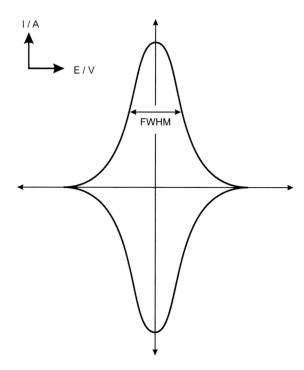

provided. Near symmetric cyclic voltammetric profiles are observed with approximately equal reduction and oxidation peak heights, characteristic of *thin layer* electrochemical behaviour. As shown in Fig. 2.45, both cathodic and anodic peak potentials remain almost unchanged over the chosen scan rate range. A plot of $\log_{10}$ (peak current) against $\log_{10}$ (scan rate) was found to linear with a slope of 0.98 (correlation coefficient of 0.999), which is very close to the expected theoretical slope of 1 for thin layer voltammetry as predicted by Eq. (2.78) [24].

In real situations, the absorbed species may be weakly or strongly absorbed. In these contexts, one usually refers to the reactants that are adsorbed, but four scenarios can be encountered, as elegantly shown in Fig. 2.46, each giving rise to unique and intriguing voltammetry.

Of note is that in the case of a strongly absorbed reactant (Fig. 2.46d) there is a pre-peak before the solution phase voltammetric peak while in the case that the product is strongly adsorbed, the adsorption wave is seen following the solution phase peak (Fig. 2.46c). The effect of varying the voltammetric scan rate can be highly illuminating, as shown in Fig. 2.47 for the case of a strongly adsorbed product, where at slow scan rates (curve A) the adsorption wave is large relative to the first diffusional peak. As the scan rate is increased, the current of the adsorption peak decreases in magnitude while the diffusional peak current increases. At very high scan rates the adsorption wave is absent (curve D) [25].

There is also another scenario which can give rise to unique voltammetry. In the context of studying new materials, such as carbon nanotubes and indeed graphene

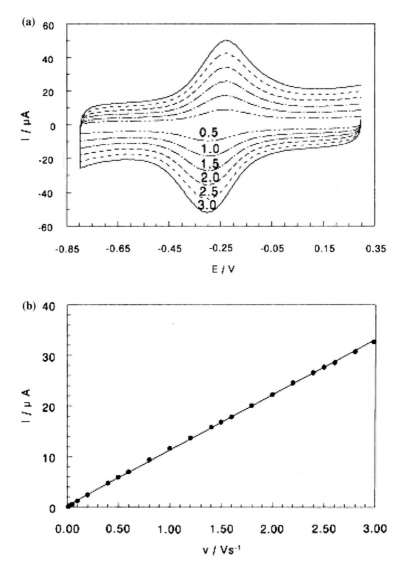

Fig. 2.45 Cyclic voltammograms **a** for Hb-DMPC films in pH 5.5 buffer at different scan rates $(V\ s^{-1})$; **b** influence of scan rate on reduction peak current. Reproduced from Ref. [24] with permission from Elsevier

which is the focus of this *handbook*, researchers usually disperse their chosen nanotubes into a non-aqueous solvent and put aliquots onto the working electrode of their choice. This modified surface is allowed to dry to enable the solvent to evaporate leaving the nanotubes immobilised upon the electrode surface, which is now ready to be electrochemically explored (the so called drop-coating method). This modified nanotube electrode surface is shown in Fig. 2.48. It has been shown

Fig. 2.46 Voltammetry with adsorption of reactants and products. **a** reactant adsorbed weakly; **b** product adsorbed weakly; **c** product adsorbed strongly; **d** reactant adsorbed strongly. *Dashed lines* are for response in the absence of adsorption. Reproduced with permission from Ref. [25], copyright 1967 The American Chemical Society

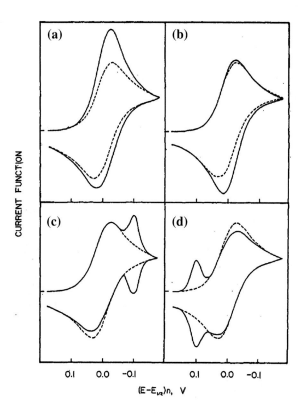

that the nanotube modified electrode exhibits a porous surface where 'pockets' of the electroactive species are trapped in-between multiple layers of nanotubes and the trapped species act akin to that of a *thin layer* cell [26]. The porous nanotube layer has a large surface area and the electrode is thought to be in contact with a finite, 'thin-layer' of solution (the species is trapped within the nanotube structure). In this case a mixture of diffusional regimes exists.

Figure 2.49 shows the voltammetry that will be observed as more nanotube material is immobilised upon the electrode surface, where there is an apparent improvement in the voltammetric peak height and a reduction in the potentials to lower values. Such a response has been assumed to be due to the electro-catalytic nature of the nanotubes themselves rather than a simple change in mass transport.

As thin-layer dominates, the ΔE_P changes from diffusional to that of thin-layer such that the peak-to-peak separation decreases giving the misleading impression that a material with fast electron transfer properties is giving rise to the response and hence misinterpretation can arise. Care also needs to be taken when adsorbing species are being explored as this will also give rise to thin layer type voltammetry [27]. Indeed the distinction between thin-layer diffusion and adsorption effects is not easy to make, especially in cases where the adsorption is rapidly reversible. Where there is slow adsorption (and desorption) kinetics, the presence or absence

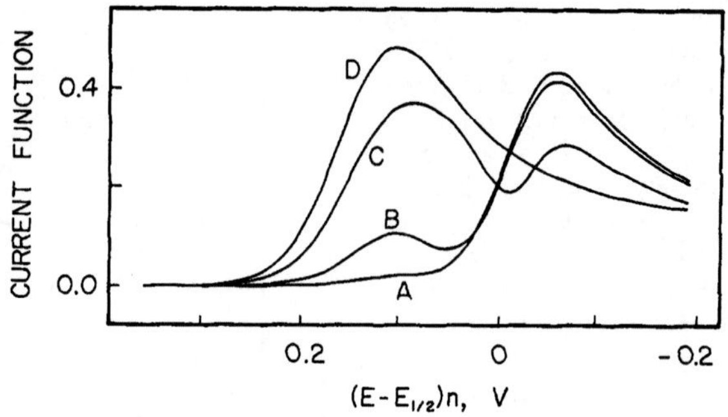

Fig. 2.47 Voltammetric response at scan rates of 1 (*A*), 25 (*B*) 625 (*C*) and 2500 (*D*) Vs^{-1} for a product strongly adsorbed. Reproduced with permission from Ref. [25], copyright 1967 The American Chemical Society

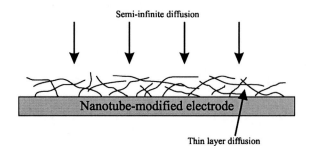

Fig. 2.48 Schematic representation of the two types of diffusion that contribute to the observed current at a highly porous CNT modified electrode. Reproduced from Ref. [26] with permission from Elsevier

Fig. 2.49 Overlaid voltammograms recorded at 100 mVs^{-1} of 1 mM dopamine at a glassy carbon electrode modified with 0 μg (*solid line*), 0.4 μg (*dashed line*) and 2.0 μg (*dotted line*) MWCNTs. Reproduced from Ref. [26] with permission from Elsevier

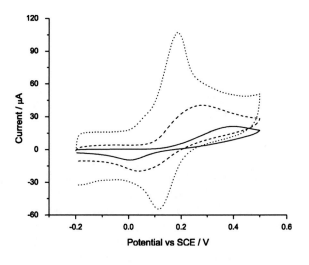

of "memory effects" can be useful. If it is possible to transfer the electrode, after exposure to the target solution, to a fresh electrolyte containing no analyte, then adsorption effects can be inferred if voltammetric signals are retained or if signals increase steadily over a period of time [27].

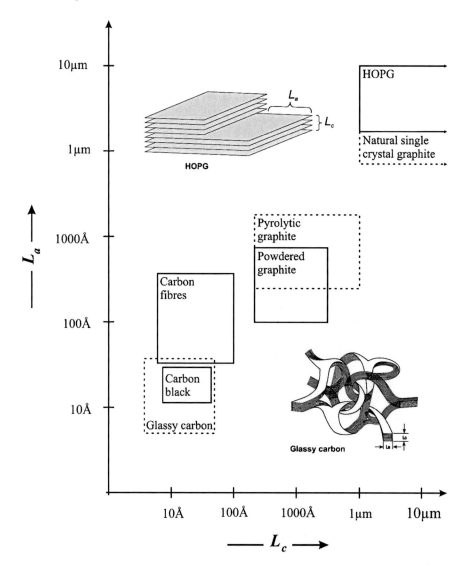

Fig. 2.50 The approximate ranges of L_a and L_c values for various sp^2 carbon materials. Note, there is large variation of L_a and L_c with sample history and thus the values shown should be considered representative, yet approximate. A schematic representation of the L_a and L_c microcrystalline characteristics of HOPG and glassy carbon is also shown

The above chapter is designed to give novices an insight into interfacial electrochemistry which can be used in interpret data presented in later chapters. Readers are directed to the following texts for further information on electrochemistry [6, 28].

2.12 Electrode Materials

There are a whole host of commercially available working electrodes which utilise a large variety of graphite products, such as amorphous carbon, glassy carbon, carbon black, carbon fibres, powdered graphite, pyrolytic graphite (PG) and highly ordered pyrolytic graphite (HOPG), each with different chemical and physical properties. The key structural factor that leads to such an assortment of different materials is the average graphite micro crystallites size (also known as the lateral grain size), L_a, which is effectively the average size of the hexagonal lattices that make up the macro structure. In principle, this can range from being infinitely large, as in the case of a macro single crystal of graphite, to the size of a benzene molecule; approximately 3 Å. In practice, the smallest L_a values are found in amorphous carbon, glassy carbon and carbon black and can be as low as 10 Å. Carbon fibres and pyrolytic graphite are intermediates in the range, with L_a values of *ca*. 100 Å and 1000 Å respectively. The assortment of materials are compared in Fig. 2.50 which reveals that the largest graphite monocrystals are found in high quality (ZYA and SPI 1 grade) HOPG, which can be 1–10 μm in size.

Regions where individual graphite monocrystals meet each other (i.e. grain boundaries) are poorly defined and when exposed result in surface defects. In the case of pyrolytic graphite, the individual graphite crystallites lie along the same axis making it possible to obtain carbon surfaces with significantly less defects. This is especially true for HOPG where the large lateral grain size can result in a well-defined surface with values of defect coverage as low as 0.2 % [29].

In Chap. 3 we consider the electrochemistry of graphene and the attempts made to understand this unique material.

References

1. T.J. Davies, C.E. Banks, R.G. Compton, J. Solid State Electrochem. **9**, 797–808 (2005)
2. H. Matsuda, Y. Ayabe, Z. Elektrochem. **59**, 494–503 (1955)
3. R.S. Nicholson, Anal. Chem. **37**, 1351–1355 (1965)
4. R.J. Klingler, J.K. Kochi, J. Phys. Chem. **85**, 1731–1741 (1981)
5. I. Lavagnini, R. Antiochia, F. Magno, Electroanalysis **16**, 505–506 (2004)
6. A.J. Bard, L.R. Faulkner, *Electrochemical Methods: Fundamentals and Applications*, 2nd edn. (Wiley, New York, 2001)
7. K. Stulik, C. Amatore, K. Holub, V. Marecek, W. Kutner, Pure Appl. Chem. **72**, 1483–1492 (2000)
8. A.C. Testa, W.H. Reinmuth, Anal. Chem. **33**, 1320–1324 (1961)
9. J.-A. Ni, H.-X. Ju, H.-Y. Chen, D. Leech, Anal. Chim. Acta **378**, 151–157 (1999)
10. R.G. Compton, C.E. Banks, *Understanding Voltammetry* (World Scientific, Singapore, 2007)
11. V.S. Vasantha, S.-M. Chen, Electrochim. Acta **52**, 665–674 (2006)
12. D. Shoup, A. Szabo, J. Electroanal. Chem. Interfacial Electrochem. **140**, 237–245 (1982)
13. M.Y. Abyaneh, J. Electroanal. Chem. **530**, 82–88 (2002)
14. M.Y. Abyaneh, M. Fleischmann, J. Electroanal. Chem. **530**, 89–95 (2002)
15. M.Y. Abyaneh, J. Electroanal. Chem. **530**, 96–104 (2002)

16. R.L. Deutscher, S. Fletcher, J. Electroanal. Chem. Interfacial Electrochem. **239**, 17–54 (1988)
17. R.L. Deutscher, S. Fletcher, J. Electroanal. Chem. Interfacial Electrochem. **277**, 1–18 (1990)
18. C.R. Wilke, P. Chang, Am. Inst. Chem. Eng. J. **1**, 264–270 (1955)
19. S. Shahrokhian, E. Asadian, Electrochim. Acta **55**, 666–672 (2010)
20. A. Molina, E. Laborda, F. Martínez-Ortiz, D.F. Bradley, D.J. Schiffrin, R.G. Compton, J. Electroanal. Chem. **659**, 12–24 (2011)
21. A.A. Barros, J.A. Rodrigues, P.J. Almeida, P.G. Rodrigues, A.G. Fogg, Anal. Chim. Acta **385**, 315–323 (1999)
22. J. Wang, J. Lu, U.A. Kirgoz, S.B. Hocevar, B. Ogorevc, Anal. Chim. Acta **434**, 29–34 (2001)
23. I. Švancara, C. Prior, S.B. Hočevar, J. Wang, Electroanalysis **22**, 1405–1420 (2010)
24. J. Yang, N. Hu, Bioelectrochem. Bioenerg. **48**, 117–127 (1999)
25. R.H. Wopschall, I. Shain, Anal. Chem. **39**, 1514–1527 (1967)
26. I. Streeter, G. G. Wildgoose, L. Shao, R. G. Comptonm, Sens. Actuators, B **133**, 462–466 (2008)
27. M. C. Henstridge, E. J. F. Dickinson, M. Aslanoglu, C. Batchelor-McAuley, R. G. Compton, Sens. Actuators, B **145**, 417–427 (2010)
28. J. Wang, *Analytical Electrochemistry*, 2nd edn. (Wiley-VCH, New York, 2000)
29. C. E. Banks, T. J. Davies, G. G. Wildgoose, R. G. Compton, Chem. Commun. 2005, 829–841
30. A.K. Yagati, T. Lee, J. Min, J.-W. Choi, Colloids Surf. B **92**, 161–167 (2012)
31. Y. Wang, E.I. Rogers, R.G. Compton, J. Electroanal. Chem. **648**, 15–19 (2010)
32. X. Cai, G. Rivas, P.A.M. Farias, H. Shiraishi, J. Wang, M. Fojta, E. Paleček, Bioelectrochem. Bioenerg. **40**, 41–47 (1996)
33. J. Wang, J. Lu, S.B. Hocevar, P.A.M. Farias, Anal. Chem. **72**, 3218–3222 (2000)
34. J. Tomeš, Collect. Czech. Chem. Commun. **9**, 12–21 (1937)

Chapter 3
The Electrochemistry of Graphene

When the seagulls follow the trawler, it's because they think
sardines will be thrown into the sea. Thank you very much.

King Eric (Cantona)

In this chapter we overview recent developments made by researchers to fundamentally understand the electrochemical behaviour of graphene as an electrode
material. However, before considering graphene, it is insightful to first overview
graphite and other graphitic surfaces, where a significant amount of information
has been gathered over many decades of research, which can be built upon and
applied to developing insights into *graphene electrochemistry*.

3.1 Fundamental Electrochemistry of Graphite

Carbon based electrode materials have long been utilised within electrochemistry
and have out-performed the traditional noble metals in many significant areas,
resulting in them being at the forefront of innovation in this field [1]. This diverse and
sustained success is due to carbons structural polymorphism, chemical stability, low
cost, wide operable potential windows, relative inert electrochemistry, rich surface
chemistry and electro-catalytic activities for a variety of redox reactions [1, 2].

Graphite surfaces are heterogeneous (anisotropic) in nature, with the overall
chemical and electrochemical reactivity differing greatly between two distinct
structural contributions which are fundamental to the behaviour of graphitic
electrodes, namely the edge and basal planes [1]. As mentioned in Chap. 2, the
intraplanar (L_a, or *basal plane*) and interplanar (L_c, or *edge plane*) microcrystallic
values define distinct structural characteristics of carbon materials (see for
example Fig. 3.3), with Highly Ordered Pyrolytic Graphite (HOPG) exhibiting the
largest graphite monocrystals; which are found in high quality (ZYA and SPI-1
grade) HOPG. Pyrolytic graphite is a graphitic material with a high degree of
preferred crystallographic orientation of the c-axes perpendicular to the surface
of the substrate (see Fig. 3.1) and is obtained by graphitisation heat treatment of
pyrolytic carbon or by Chemical Vapour Deposition (CVD) at extremely high
temperatures ($\sim 2,500$ °K). The hot working of pyrolytic graphite by annealing
under compressive stress at high temperatures produces HOPG. The crystal
structure of HOPG is shown in Fig. 3.1 which is characterised by an arrangement

D. A. C. Brownson and C. E. Banks, *The Handbook of Graphene Electrochemistry*,
DOI: 10.1007/978-1-4471-6428-9_3, © Springer-Verlag London Ltd. 2014

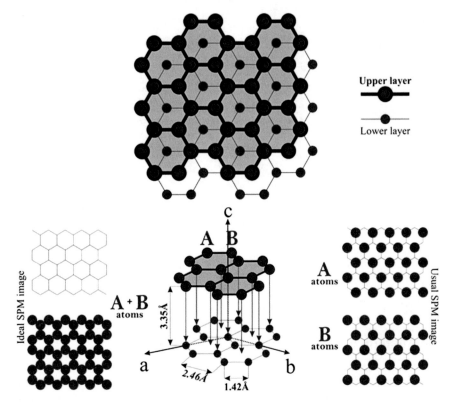

Fig. 3.1 A schematic representation of the structure of a bulk *hexagonal* graphite crystal showing the bulk unit cell. *Side insets*: *Top* view of the basal plane of graphite and a schematic representation of the surface structure (carbon atoms) of graphite, where every other atom is enhanced (*right-side inset*) and viewed under ideal conditions, and where every single atom is seen (*left-side inset*). Figure reproduced from Ref. [3]

of carbon atoms in stacked parallel layers, where the graphite structure is described by the alternate succession of these identical staked planes. Carbon atoms within a single plane of graphite have a stronger interaction than with those from adjacent planes (which explains the cleaving behaviour of graphite). Note that a single-atom thick form of carbon is known as graphene, where the lattice consists of two equivalent interpenetrating triangular carbon sub-lattices denoted A and B (see Fig. 3.1) where each one contains a half of the carbon atoms. Each atom within a single plane has three nearest neighbours: the sites of one sub-lattice (A—marked by the red layer in Fig. 3.1) are at the centres of triangles defined by three nearest neighbours of the other one (B—marked by the blue layer in Fig. 3.1). The lattice of graphene has two carbon atoms, designated A and B, per unit cell, and is invariant under 120° rotation around any lattice site.

The term "mosaic spread" is used to characterise the quality of HOPG which is performed via X-ray crystallography with CuKα radiation by measuring

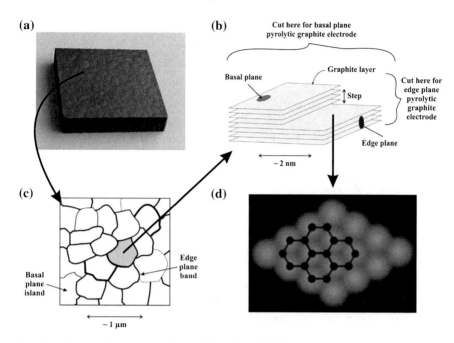

Fig. 3.2 a Image of a commercially available slab of HOPG. **b** Schematic representation of the side on view of a HOPG surface, highlighting its basal plane and edge plane like- sites/defects which exhibit contrasting behaviours in terms of electrochemical activity, where electron transfer kinetics of the latter are overwhelmingly dominant over that of the former which in comparison are relatively (electrochemically) inert. **c** A schematic representation of a HOPG surface showing the discrete basal plane and edge plane islands. **d** A typical STM image of a HOPG surface with the corresponding fragment of the graphene structure superimposed

(in degrees) a Full Width at Half Maximum (FWHM) of the rocking curve. Disorder in the HOPG produces broadening of the (002) diffraction peak where the more disorder, the wider the peak becomes. The measured value of the mosaic spread depends not only on crystal quality, but also on the energy and the cross section of the reflected beam. Figure 3.2a depicts a picture of a commercially obtainable HOPG 'slab' which in this case is 'Grade SPI-1', which has the tightest mosaic spread of $0.4°$ ($\pm 0.1°$) demonstrating outstanding crystalline perfection. Note that the ZYB, ZYD and ZYH grade HOPG results in mosaic spread values of $0.8°$ ($\pm 0.2°$), $1.5°$ ($\pm 0.3°$) and $3.5°$ ($\pm 0.5°$) respectively.

Shown in Fig. 3.2 is a top-down schematic representation of the HOPG surface, which depicts the discrete edge plane and basal plane islands, and a side on view highlighting the edge plane and basal plane like- sites/defects which are defined by the quality of the chosen HOPG. Also shown in Fig. 3.2d is a Scanning Tunnelling Microscopy (STM) image of a HOPG surface, highlighting the hexagonal crystal structure. Note that in terms of the electrochemical performance of graphitic materials, it has been deduced (see Sect. 3.1.2) that the electrochemical activity of edge and basal planes is distinct such that electrochemical reactions on the edge

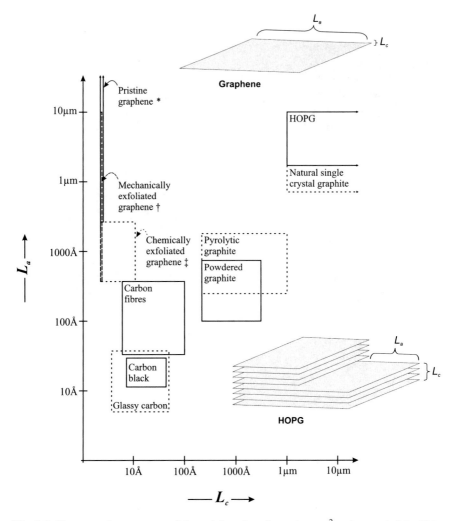

Fig. 3.3 The approximate ranges of L_a and L_c values for various sp^2 carbon materials. Note, there is large variation of L_a and L_c with sample history and thus the values shown should be considered representative, yet approximate. *: Pristine graphene; commercially available from 'Graphene Supermarket', produced via a substrate-free gas-phase synthesis method. [7, 8] ‡: Chemically exfoliated graphene; commercially available from 'NanoIntegris', produced via a surfactant intercalation process—note that this range is also representative of graphene produced through other chemical exfoliation routes such as the reduction of GO. [9, 10] †: Mechanically exfoliated graphene was fabricated through the so-called 'scotch tape method'. Note that graphene synthesised via CVD has been excluded given that crystal size and quality are large variables through this route, however single graphene crystals with dimensions of up to 0.5 mm have been reported. [11, 12] A schematic representation of the L_a and L_c microcrystalline characteristics of graphene and *HOPG* is also shown. Reproduced from Ref. [13] with permission from The Royal Society of Chemistry

plane like- sites/defects are anomalously faster (exhibit greater reactivity) over that of the basal planes [4–6]. In terms of relating this to other carbon allotropes, Fig. 3.3 shows the range of L_a and L_c values for a collection of other graphitic forms where it is evident that HOPG has L_a and L_c values exceeding 1 μm while polycrystalline graphite has values from 10 to 100 nm and carbon from 1 to 10 nm. Graphene, which is readily obtainable from a range of commercial suppliers is also included, highlighting the variation in structure that can be obtained, which is of course dependant on the fabrication methodology; with L_a values for graphene ranging from below 50 up to 3,000 nm and larger, and of course true (monolayer) graphene possesses an L_c value of 0.34 nm.

3.1.1 The Electronic Properties (DOS) of Graphitic Materials

An important parameter of an electrode material is its electronic properties, namely, the Density Of electronic States (DOS) which varies greatly on the different forms of graphite. Gold typically has a DOS of 0.28 states atom^{-1} eV^{-1} with the high conductivity of gold arising from the combination of a high proportion of atomic orbitals to form bands with a high density of electronic states [1]. For a given electrode material, a higher DOS increases the possibility that an electron of the correct energy is available for the electrode to transfer to an electroactive species; the heterogeneous electron transfer rate is thus dependent on the DOS of the electrode material [1]. HOPG has a DOS which overall is lower than that of metal, but is particularly low near the Fermi level and has been reported to have a minimum DOS of around 0.0022 states atom^{-1} eV^{-1}, which is about 0.8 % that of gold [1].

The DOS at graphitic materials can be increased through disorder such that electroactive species exhibit increasing electron transfer rates but by varying amounts. In terms of outer-sphere electron transfer systems, disorder increases the rate by modifying the electronic structure of the carbon while for inner-sphere systems, specific surface interactions also contribute (see later, Box 3.1) [14]. A perfect/pristine basal surface of HOPG has no edge plane (in theory), no location for surface functional groups and there are no dangling bonds since the carbon atoms have satisfied valances [1]. When disorder is introduced, such as through mechanical roughening of the electrode surface, the surface is disturbed such that surface defects are introduced, viz edge plane sites, which increase the DOS [1]. A further extreme is the complete change of a graphitic surface to a different structural composition (L_a and L_c; see Figs. 3.2 and 3.3) towards that of Edge Plane Pyrolytic Graphite (EPPG) which has a high proportion of edge plane sites and thus improvements in electron transfer are observed [1, 13].

Electronic properties of graphitic materials are thus highly relevant and critical, where the energy-dependant densities of electronic states have major effects on electron transfer. Note that graphitic materials differ greatly in their surface chemistry, which is also critical when understanding electrochemical processes at these materials [1]. Such insights from graphitic materials can be applied for the case of graphene. In terms of the DOS for graphene, insights from HOPG (multiple layers of graphene) can be illuminating to understand its electrochemical reactivity. For a diffusional outer-sphere electron transfer process, the standard electrochemical rate constant, k^0, can be defined as [15, 16]:

$$k^o = \frac{(2\pi)^2 \rho \left(H_{DA}^0\right)^2}{\beta h (4\pi\Lambda)^{1/2}} \exp\left[-\frac{\Lambda}{4}\right] I(\theta, \Lambda) \qquad (3.1)$$

where ρ is the density of electronic states in the electrode material, H_{DA}^o is the electronic coupling matrix at the closest distance of approach, $\Lambda = (F/RT)\lambda$, where λ is the reorganisation energy, β is its associated electronic coupling attenuation coefficient, h is Planck's constant, F is the Faraday constant, R the gas constant and T the absolute temperature. $I(\theta, \Lambda)$ is an integral give by [16]:

$$I(\theta, \Lambda) = \int_{-\infty}^{\infty} \frac{\exp\left[-(\varepsilon - \theta)^2/4\Lambda\right]}{2\cosh[\varepsilon/2]} dE \qquad (3.2)$$

where $\theta = F/RT(E - E_f^0)$, and E_f^0 is the formal potential. Thus from inspection of Eq. (3.1) there is a direct relationship between the DOS and the standard electrochemical rate constant (k^o). Thus this can be interpreted for graphite as the DOS varies significantly as a function of energy with a minimum at the Fermi level [16]. For example, it has been shown that electron transfer is non-adiabatic and that the rate of electron transfer varies as a function of the applied potential, as is evident from inspection of Eq. (3.1) for outer-sphere redox systems [16].

It has been reported that the basal plane of pristine graphene has a DOS of 0 at the Fermi level, which was shown to increase with edge plane defects [17–19]. Conversely the edge plane sites on graphene nanoribbon's zigzag edge have been reported to possess a high DOS [19]. Other work has shown that depending on how the edge of graphene terminates, [20] a variable DOS is observed [21]. Thus, graphene, a single layer comprising HOPG, should in theory act similar in terms of its DOS to that observed for HOPG (see above); that is, pristine graphene with no defects should exhibit poor electrochemical behaviour and on the contrary graphene possessing a high degree of defects should exhibit improvements in the observed electrochemical rate constant.

There is a wealth of literature on graphene which reports that the edge of graphene is particularly more reactive than its side (basal plane). For example, using Raman spectroscopy Strano and co-workers [22] report the reactivity of graphene, that being single-, double-, few- and multi-layer towards electron transfer chemistries with 4-nitrobenzenediazonium tetrafluoroborate. Strano et al. [22]

Fig. 3.4 a Schematic representation of an electrochemical reaction occurring on the same electrode surface with different Butler-Volmer characteristics; and a top-down perspective (**b**)

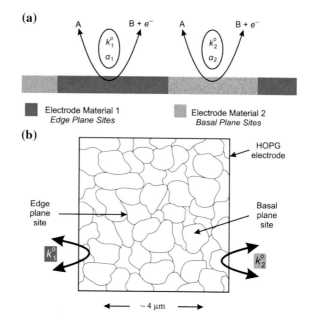

interpret their observations with consideration to the Gerischer–Marcus theory which states that the charge transfer depends on the electronic DOS of the reacting species and is not restricted to their Fermi levels only. The observed electron-transfer reaction rate ($k_{Graphene}^{OBS}$) is given by Eq. (3.3) where $W_{OX}(\lambda, E)$ is the distribution of the unoccupied redox states of the electron acceptor in solution given by Eq. (3.4). The $DOS_{Graphene(N=1/N=2)}$ is the electronic density of states of graphene for $N = 1$, and of double layer graphene for $N = 2$ and ε_{OX} is the proportionality function [22].

$$k_{Graphene}^{OBS} = \upsilon_n = \int_{E_{redox}}^{E_F^{Graphene}} \varepsilon_{OX}(E) DOS_{Graphene(N=1/N=2)}(E) W_{OX}(\lambda, E) dE \quad (3.3)$$

$$W_{OX}(\lambda, E) = \frac{1}{\sqrt{4\pi\lambda kT}} exp\left(-\frac{(\lambda - (E - E_{redox}))^2}{4\lambda kT}\right) \quad (3.4)$$

Calculations presented by Strano et al. suggest that double layer graphene is almost 1.6 times more reactive than single layer graphene [22]. Thus based on the DOS, it is clear that double layer graphene (or further graphitic structures consisting of multiple graphene layers) is more reactive than single layer graphene, which has clear implications for graphene as an electrode material; it is this we explore in more detail in Sect. 3.2.

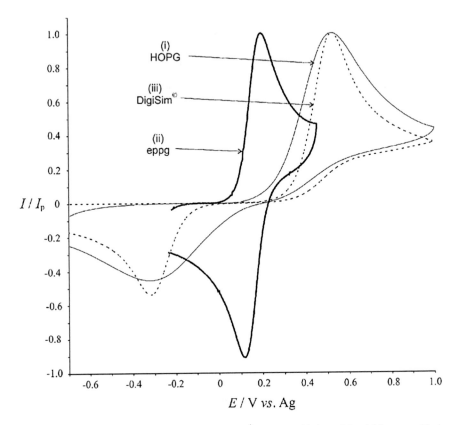

Fig. 3.5 Cyclic voltammograms recorded at 1 Vs^{-1} for the oxidation of 1 mM ferrocyanide in 1 M KCl at a basal plane *HOPG* electrode and an *EPPG* electrode. The *dashed line* voltammogram is the simulated fit using linear diffusion only (DigiSim$^{(R)}$). Reproduced from Ref. [23] with permission from The Royal Society of Chemistry

3.1.2 Electrochemistry of Heterogeneous Graphitic Surfaces

The electrochemical characteristics and reactivity of HOPG has been fully understood by Compton and co-workers, [5] who have shown convincing evidence that edge plane sites/defects are the predominant origin of electrochemical activity. Figure 3.4a shows a schematic representation of the heterogeneous HOPG surface (see Fig. 3.2b and c) which has the two distinctive structural contributions, namely edge plane and basal plane sites, each with their own electrochemical activity and thus differing Butler-Volmer terms, k^o and α.

Using a simple redox couple, Fig. 3.5 depicts the voltammetry obtained when using either a Basal Plane Pyrolytic Graphite (BPPG) (i) or (ii) an EPPG electrode of HOPG, and the responses are compared with numerical simulations (iii) assuming linear diffusion only, in that, all parts of the electrode surface are uniformly (incorrectly) electrochemically active. Two features of Fig. 3.5 are to be

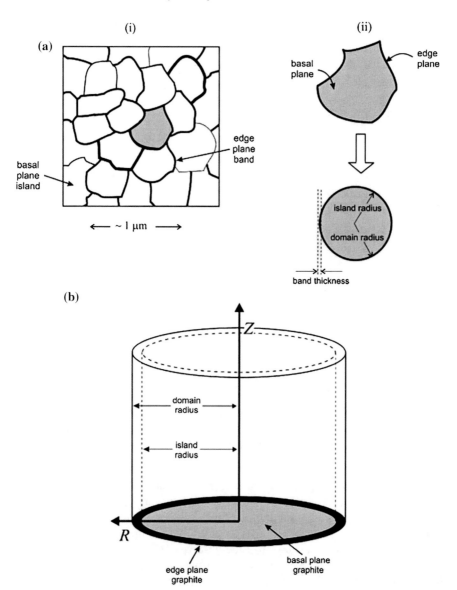

Fig. 3.6 Schematic diagrams showing: **a** (*i*) the overhead view of a section of the basal plane HOPG surface and (*ii*) the approximation of each island/band combination as a partially covered *circular disc* of the same area; **b** the resulting diffusion domain from the approximation in (**a**) (*ii*) and the cylindrical coordinate system employed. Reproduced from Ref. [23] with permission from The Royal Society of Chemistry. Note that the island radius is termed R_b and the domain radius is R_0

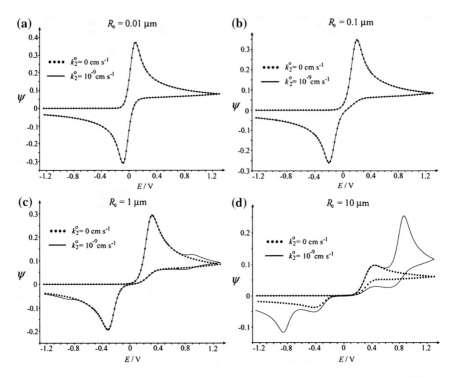

Fig. 3.7 *Solid curves* are simulated dimensionless current cyclic voltammograms for diffusion domains where $D = 6.1 \times 10^{-6}$ cm^2 s^{-1}, $k_1^o = k_{edge}^o = 0.022$ cm s^{-1}, $k_2^o = k_{basal}^o = 10^{-9}$ cm s^{-1}, $\upsilon = 1$ V s^{-1}, the band thickness is 1.005 nm and the domain radius is **a** 0.01 µm **b** 0.1 µm **c** 1 µm and **d** 10 µm. Overlaid in each section are the simulated inert equivalents (*dotted curves*), i.e., $k_2^o = k_{basal}^o = 0$ cm s^{-1}. Reprinted from Ref. [5] with permission from Elsevier

noted: (1) there is a significant increase in the peak-to-peak separation, ΔE_P, observed for (iii) over the EPPG voltammetric response (ii); (2) the fit to the 'linear diffusion' only (iii) simulation is not fully satisfactory, especially in the return scan where a significantly lower back peak (current) is observed than expected [23]. It has been shown that the observed voltammetric signature (i) can be correctly and quantitatively simulated through considering the HOPG surface (as shown in Figs. 3.2 and 3.4) to be a heterogeneous surface consisting of edge plane nano bands which have been concluded to be exclusively the sites of electrocatalysis whereas the basal plane 'islands' are electro-catalytically inert [5].

Figure 3.6 depicts how the HOPG surface has been simulated using numerical simulation via the diffusion domain approach, where each basal plane island and the surrounding edge-plane band is considered as a circular disc of edge-plane graphite partially (or almost completely) covered with basal plane graphite, such that the areas of edge and basal plane are consistent. Since the island and band are surrounded by other island/band combinations, little or zero net flux of electroactive species will pass from one island to its neighbour [5, 23].

Fig. 3.8 **a** The surface is split into a series of identical domains (unit cells), namely band islands. **b** Schematic showing the difference between diffusion to macro- (*i*) and micro- (*ii*) scale electrode systems. The *darker* area represents the island (r_{band}) with the faster kinetics. Reproduced from Ref. [24] with permission from The Royal Society of Chemistry

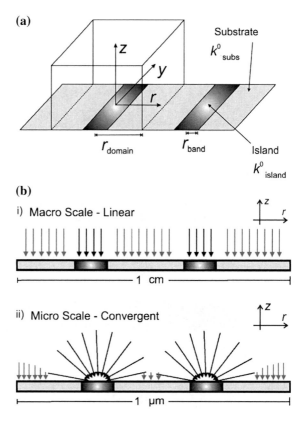

The circular discs are treated as independent entities with cylindrical walls through which no net flux can pass. These unit cells are better known as diffusion domains and are illustrated in Fig. 3.6 where the two electrode materials (edge plane and basal plane) are highlighted. The voltammetric response of the whole HOPG electrode is therefore the sum of that for every domain on the electrode surface. Also shown is a single diffusion domain unit cell and the cylindrical polar coordinate system employed where interacting cylindrical units of radius R_0 are centred around a circular block of radius R_b, where the fractional coverage of the domain, $\theta = R_b^2/R_0^2$ such that the surface areas of the basal sites and edge sites are given by $(1 - \theta)\pi R_0^2$ and $\theta\pi R_0^2$ respectively, allowing the effect of varying the edge sites while keeping the surface coverage constant. The island radius is termed as R_b and R_0 is the domain radius which includes the width of the edge plane site/band. As is evident from Fig. 3.7, the ΔE_P of the edge plane nano band signal depends strongly on the edge plane coverage, and the domain size has little or no influence on the observed voltammetry of the three smaller domains due to the depleting effect of non-linear diffusion which becomes less relevant as the domain sizes increase. Note that the maximum lateral grain size of HOPG is 1–10 µm resulting in a maximum R_0 of ∼0.5–5 µm, the edge plane coverage is such that the basal plane

(a)

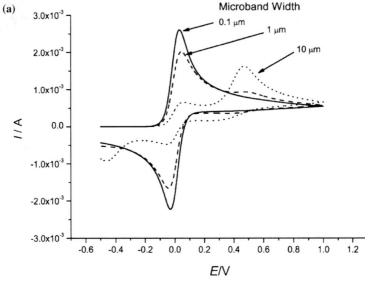

(b)

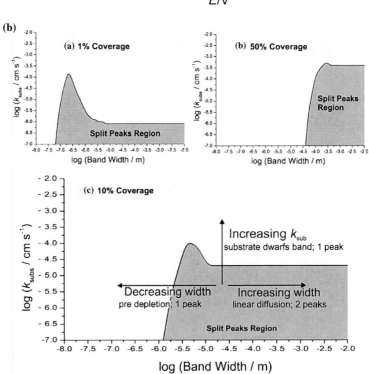

◀ **Fig. 3.9** **a** Voltammetry of a one-electron transfer process at an electrochemically heterogeneous electrode consisting of an array of microbands ($k^o = 10$ cm s^{-1}) distributed over a substrate material ($k^o = 10^{-6}$ cm s^{-1}) of area 1 mm^2 and a surface coverage of the bands of 10 % at a scan rate of 0.1 Vs^{-1}. The diffusion coefficient of all species is 10^{-5} cm^2 s^{-1} with an initial concentration of 10 mM. The voltammetry transitions from 1 peak to 2 peaks as the width of the band (labelled) is increased. **b** Schematics showing the region of the 'Band Width'-'Substrate rate constant' space for which there are two peaks in the forward sweep of a cyclic voltammogram at band surface coverages of (*a*) 1 % (*b*) 50 % and (*c*) 10 %. Scan rate = 0.1 Vs^{-1}; diffusion coefficient = 10^{-5} cm^2 s^{-1}; island rate constant $k^o_{band} = 10$ cm s^{-1}. Reproduced from Ref. [24] with permission from The Royal Society of Chemistry

is effectively inert [23] and the HOPG response can be assigned to nano bands of edge plane graphite with the basal plane islands having no contribution.

Further work from the Compton group has explored the 'double peak concept', [24] modelling a HOPG surface as an array of microbands; the unit cell is shown in Fig. 3.8a while Fig. 3.9 depicts the response of an electrochemically heterogeneous surface highlighting the effect of microband width along with the domain coordinates utilised where the fractional coverage of the surface covered is given by: $\theta_{band} = r_{band}/r_{domain}$. Figure 3.9a shows that as the width of the band is increased the diffusion profile changes from being largely convergent, as shown schematically in Fig. 3.8b, to that of linear which is seen as one peak becoming two and a decrease in the peak current is also evident; note that Chap. 2 considered the case of mass transport. The depletion of the electroactive species above the electrochemically slower substrate proceeds to a greater extent so the substrate has less of an influence on the diffusion of the electroactive species and thus less of an influence on the observed voltammetry [24]. The depletion, known as the diffusion layer, is given by Eq. (2.32) where for voltammetry t is replaced by: '$\Delta E/v$' where ΔE is the potential range over which electrolysis occurs and t (referred to as t_{peak}: below) is the time taken to sweep the potential from its initial value to the point where the current reaches a maxima. It has been shown that [24]:

$$\sqrt{2Dt_{peak}} \gg r_{sep} \tag{3.5}$$

where

$$r_{sep} = \frac{1}{2}\left(r_{domain} - r_{band}\right)$$

where r_{band} is the width of the edge plane site and r_{domain} accounts for the edge plane site *plus* the basal plane site (see Fig. 3.8a). In this case, the inter-band separation is small compared to the extent of diffusion parallel to the electrode surface and only one peak will be observed in the voltammetry. Figure 3.9b shows the effect of the band width upon k^o_{subs} and in which region split peaks will be observed. In the case that there is a large domain width, $\sqrt{2Dt_{peak}} \ll r_{sep}$, such that the voltammetry will be a superposition of the voltammetry of the band and substrate in isolation where diffusion to each will be linear in nature.

If the heterogeneous rate constants on the two electrode surfaces are similar, two peaks will be observed arising at similar potentials which will merge into one larger peak. If $k_{band}^o \gg k_{subs}^o$ (i.e. $k_{edge}^o \gg k_{basal}^o$) two peaks will be observed if the k_{subs}^o has measurable activity; however it has been shown that this is not the case and only the k_{band}^o is active, or sometimes reported as anomalously faster over that of k_{subs}^o [4].

The rate of electron transfer for basal plane sites has been reported to correspond to $\sim 10^{-9}$ cm s^{-1} for the oxidation of ferrocyanide and is considered to be possibly even zero [4–6]. *How does one know that this is actually correct?* As shown in Fig. 3.9a, a strangely distorted voltammogram would be observed in the limit of very low defect density [23]. Due to the fact that two peaks have never been observed experimentally, it is generally accepted that edge plane electron transfer kinetics are anomalously faster over that of basal plane; the latter is sometimes referred to as being inert [5, 6, 23]. Interested readers are directed to the elegant work of Davies et al. and Ward et al. to further appreciate this work [5, 24].

Further evidence on the role of edge plane sites *versus* basal plane sites has been reported [4] by the selective blocking of the basal plane sites of HOPG with a polymer whilst the edge plane sites were left exposed. Identical voltammetric behaviour was observed with this modified surface when compared to that of the initial bare electrode and with numerical simulations, confirming the edge planes to be the sites of electrochemical activity; Fig. 3.10 depicts how this was achieved.

During each stage (as shown in Fig. 3.10) each surface was voltammetrically examined and the corresponding voltammograms are depicted in Fig. 3.11. As shown in Fig. 3.11 it is evident that the final stage is nearly identical with that of stage 1 (a freshly produced HOPG surface) despite the basal plane sites being covered. The small deviation is reported to be due to the treatments that the electrode has undergone and a slight loss in activity of the edge plan sites. In the case of the modified electrode, Fig. 3.11d, only the edge-plane steps located along the bottom of the nanotrenches are exposed to the solution such that an array of nanobands have been created. This work nicely demonstrates that the cyclic voltammetric response of a basal-plane HOPG electrode (BPPG) is solely due to the edge-plane defects present, no matter how small their coverage may be, and that the basal-plane graphite terraces have no influence on the voltammetry and are effectively inert; hence blocking the basal-plane sites results in no overall change to the observed voltammetry.

Last, it is important to note that researchers will (and have already done so) dispute the extensive literature reported above. As such it has been reported that under certain (limited) conditions the basal plane sites have measurable electrochemical activity [25–27]. Using elaborate Scanning Electrochemical Cell Microscopy (SECM) it has been reported that the basal plane sites of *freshly exposed* HOPG display considerable electroactivity which, interestingly, is time dependant, in that exposure to air for less than one hour after cleaving leads to a decrease in the observed electron transfer rates at the basal surface [27]. Such work is highly fascinating and studies into this time-dependent surface effect are, at the

Fig. 3.10 Initially a HOPG surface is cleaved to produce a fresh surface (*stage 1*). In *stage 2*, MoO_2 nanowires are formed exclusively along the edge plane sites. In *stage 3*, the basal plane sites are covered by the electrochemical reduction of 4-nitrobenzenediazonium cations. *Stage 4* then involves exposing the edge plane sites by dissolution of MoO_2 in HCl. Reproduced from Ref. [4] with permission from Wiley

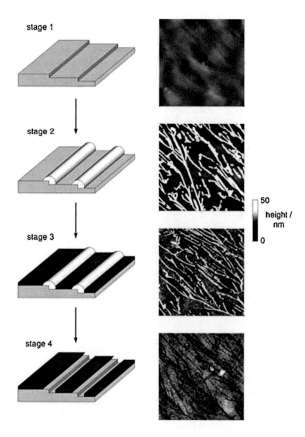

time of writing this *Handbook*, underway [27]. However, ultimately this means that over the lifetime of an experiment the observed electroactivity of the freshly cleaved basal plane sites of HOPG becomes negligible as previously reported [4–6]. Furthermore, an important challenge that has not been realised in Ref. [27] is the correlation of this local microscopic result to that of the well documented macroscopic response of a HOPG electrode; [28] i.e. if such 'pristine' HOPG surfaces, as used within Ref. [27], are used in a conventional cyclic voltammetric experiment, do the voltammograms appear fully reversible or not? If so can the pristine surfaces be reproduced by other groups? If not, then why not? [28].

3.2 Fundamental Electrochemistry of Graphene

When graphene is immobilised upon an electrode surface, as is common practice in the literature to electrically 'wire' (connect to) graphene and study its electrochemical activity, a heterogeneous electrode surface is formed. In this scenario, if we consider that a HOPG surface is utilised, Fig. 3.12 shows that four key sites

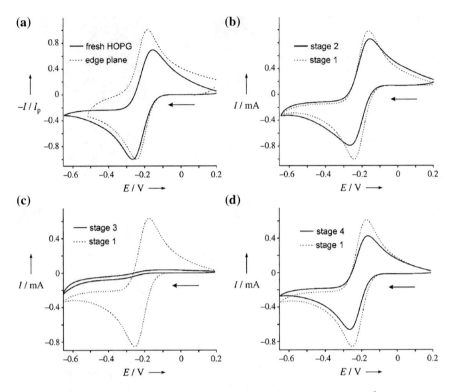

Fig. 3.11 Cyclic voltammograms for the reduction of 1.1 mM [Ru(NH3)$_6$]$^{3+}$ at a *HOPG* electrode (*vs.* SCE) after each stage of nanotrench fabrication (Fig. 3.10): **a** *stage 1*, **b** *stage 2*, **c** *stage 3*, **d** *stage 4*. The voltammograms in (**a**) were obtained from the same experiment with an EPPG electrode. Voltammograms in (**b**) through to (**d**) were obtained after *stage 1* of nanotrench fabrication. Reproduced from Ref. [4] with permission from Wiley

are evident. It is clear that in addition to the underlying HOPG electrode surface which has edge plane and basal plane sites, each with the their own electrochemical activity with different Butler-Volmer terms, k^o and α, the immobilisation of graphene with its own edge and basal plane sites (with their own k^o and α values) gives rise to an interesting situation. This scenario occurs when any carbon based electrode is utilised as the supporting electrode.

In the case of immobilising graphene upon a metallic electrode as is sometimes spuriously undertaken, such as a gold macroelectrode, there would be three key electrochemical sites, the underlying gold (k^o_{gold}, α_{gold}) and modified graphene with the contribution from edge and basal plane sites. However this is not a good situation as the underlying gold generally has (depending on the electroactive analyte) a greater electrochemical activity which dominates over the graphene (or other graphitic materials which could be employed) such that the contribution from graphene will not be solely observed or could be misinterpreted as graphene

Fig. 3.12 a A schematic representation of an electrochemical reaction occurring on a graphene modified HOPG surface exhibiting differing Butler-Volmer characteristics; and a top-down perspective (**b**). Note that the figure is obviously not to scale

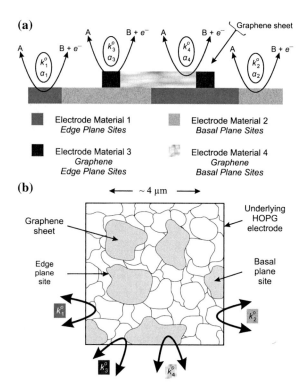

exhibiting excellent electrochemical activity if control experiments (bare/unmodified gold electrode) are not diligently undertaken.

Returning to the case of graphene as shown in Fig. 3.13 and utilising the insights discussed above from numerical simulations and experimental observations for graphitic electrodes (Sect. 3.1.2), *should graphene be considered similar to that of HOPG but as a single layer?* Given that graphite surfaces are heterogeneous (anisotropic) in nature, with the overall chemical and electrochemical reactivity differing greatly between two distinct structural contributions which are fundamental to the behaviour of graphitic electrodes, namely the edge and basal planes, graphene, a single layer that comprises HOPG, should in theory act similar in terms of its DOS to that observed for HOPG; Fig. 3.13 shows this concept.

If we assume that graphene is immobilised upon an electrode surface (and will naturally prefer to lie parallel rather than vertical due to π-π stacking) such that the underlying electrode is completely covered, we can approximate this graphene modified electrode surface to that shown in Fig. 3.6 where we have edge plane sites (viz the peripheral edge of graphene) and a basal plane (assuming for simplification that we have true graphene, that is pristine graphene with no defects across the basal surface), thus the same unit cell can be used and an approach can be taken via the diffusion domain method, which will be applicable, in that we have islands with nano edge bands of thickness which, at best, might approximate to the length of a carbon-carbon bond which is reported to be $\sim$0.142 nm in graphene. This can be

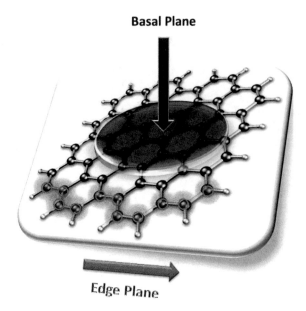

Fig. 3.13 A conceptual model depicting the structure of pristine graphene, showing the sites of electron transfer, basal and edge plane like- sites. Reproduced from Ref. [12] with permission from The Royal Society of Chemistry

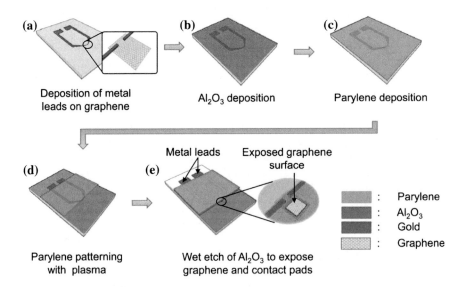

Fig. 3.14 Procedure for fabricating monolayer graphene sheets into working electrodes for electrochemical characterisation. Reprinted with permission from Ref. [29]. Copyright 2011 American Chemical Society

assumed to be constant in many different forms of graphene with the expectation that the domain radius changes (viz the L_a size) as the size of the graphene flake is either increased or decreased depending on its fabrication methodology.

If we consider two contrasting scenarios, assuming in the first instance that the edge plane has fast electron transfer activity and that the basal plane has some negligible activity ($\sim 10^{-9}$ cm s^{-1}), as is widely accepted in the case of HOPG and since graphene is simply one layer of this, the effect of the domain radius (L_a size) that will be expected can be observed from inspection of Fig. 3.7, showing that at a large domain radius two peaks might be observed. However, to date such voltammetric signatures have not yet been presented, adding weight to the inference that (as per the second scenario) the edge plane side of graphene is the electroactive site acting akin to an edge plane nano band; this is a reoccurring theme which we explore later with experimental evidence from throughout the literature. Thus in the case of modifying a HOPG surface with graphene, it is likely that the basal plane HOPG surface (BPPG), the k^o_{basal}, can be neglected such that Fig. 3.12 simplifies to two key domains, k^o_{edge} (HOPG) and k^o_{edge} (graphene), and assuming these are electrochemically similar in terms of the DOS, it is clear that edge plane sites are the key dominating factor of a graphene modified electrode.

3.2.1 Graphene as a Heterogeneous Electrode Surface

The electrochemistry of true graphene, that is, an individual monolayer crystal, has been reported by Li et al. [29] with their fabrication procedure overviewed in Fig. 3.14 where a monolayer graphene sheet is first deposited onto a SiO$_2$-coated Si substrate. In their study two types of graphene were explored, which were either fabricated via mechanical exfoliation (the so-called "scotch tape method" as covered in Chap. 1) or through CVD graphene growth. In both cases optical lithography was employed in order to be able to connect to each piece of graphene with two metal leads, as shown in Fig. 3.14a. Note that those working on the fabrication of graphene forget that one needs to somehow electrically wire and connect to the graphene sheet in order to utilise and study it!

In this work [29], after depositing the graphene sheet a 100 nm thick Al$_2$O$_3$ layer (see Fig. 3.14b), followed by a 600 nm thick parylene layer (see Fig. 3.14c), are deposited in order to isolate the metal leads from the solution such that when electrochemical experiments are performed the graphene response will be observed and the electrochemistry will not be dominated by the metal leads. This is achieved by employing an oxygen plasma to remove a region of the parylene layer above the graphene while keeping the metal leads covered (Fig. 3.14d). Finally, a window through the Al$_2$O$_3$ layer is made using a wet etch to expose a well-defined area of the graphene surface (Fig. 3.14e).

The elegant design of experimental set-up by Li et al. [29] ensures that graphene is the only electrochemically active surface that is in contact with the

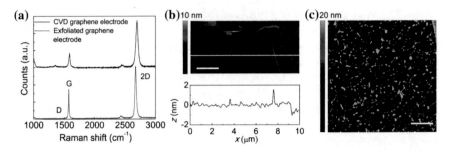

Fig. 3.15 Raman and AFM characterisation of the graphene working electrodes. **a** Raman spectra of the graphene working electrodes, after the device fabrication process was completed. **b** AFM image of the surface of a working electrode made of exfoliated graphene. A cross-sectional profile is given for the line in the top panel. Scale bar: 2 μm. **c** AFM image of the surface of a working electrode made of CVD graphene. Note that the entire area in (**c**) is within the surface of the CVD graphene electrode. Scale bar: 2 μm. Reprinted with permission from Ref. [29]. Copyright 2011 American Chemical Society

solution during electrochemical measurements. Additionally the fabrication steps were chosen to minimise the likelihood of contaminating the graphene. The maximum sizes of the exposed graphene surfaces were reported to be 15×15 μm^2 for the mechanically exfoliated graphene and 0.38×0.50 mm^2 for the CVD grown graphene since CVD graphene can be formed in much larger sheets than exfoliated graphene. In the final step, vacuum annealing at 350 °C was used in an attempt to remove organic residuals which might have remained on the graphene surface after processing [29].

Figure 3.15 shows Raman and AFM characterisation of the fabricated graphene electrodes where a symmetric single peak is observed for the 2D band, the intensity of which is significantly higher than that of the G peak confirming that high-quality single graphene layers have been obtained from the fabrication methodologies. Additionally note that a small D peak is also observed (for CVD) indicating a less pristine layer, as is generally the case for such samples. Electrodes made via this route were found to be more disordered including large wrinkles, particulates, and domain-like structures following the CVD growth and transfer process (Fig. 3.15c). For electrodes made via mechanical exfoliation (in this case), no observable D peak was evident at 1,350 cm^{-1}, indicating that the graphene sheet is clean and (at the resolution limit of micro-Raman) defect-free [29]. The step height of the graphene layer with respect to the SiO$_2$ substrate was $\sim$0.8 nm, in good agreement with the known value for clean graphene (0.5–1 nm).

Electrochemical experiments revealed a microelectrode response (steady-state current; see Chap. 2) at both the graphene electrodes, with the effective area of the graphene electrode deduced from the following equation [29]:

$$A_{eff} = \pi \left(\frac{i_{ss}}{4nFDC} \right)^2 \tag{3.6}$$

where A_{eff} is the effective area of the microelectrode/graphene electrode and i_{ss} is the steady-state current. The A_{eff} of the graphene electrodes was estimated to correspond to $117 \pm 8 \ \mu m^2$, which is in good agreement with Atomic Force Microscopy (AFM) (found to correspond to $\sim 130 \ \mu m^2$). [29] The standard electrochemical rate constant was deduced for FcMeOH to be $\sim 0.5 \ cm \ s^{-1}$ indicating an electrode with fast electron transfer kinetics [29]. The authors' infer that improvements in the electron transfer kinetics (observed when contrasted to the basal plane of HOPG) are due to corrugations in the graphene sheet, [29] or could arise from edge plane like- sites/defects across the basal plane surface of the graphene in addition to exposed edges acting like ultra-microelectrodes, with the sigmoidal voltammetry arising from the change in mass transport—the observations and inferences are highly fascinating, indicating why graphene is being fundamentally studied.

The electrochemistry of individual single and double layered graphene crystals has also been reported by Dryfe and co-workers [30] who performed time consuming experiments producing single mono-, bi- and multi- layer graphene crystals viz mechanical exfoliation ('scotch tape method') after which the authors electrically connected their graphene samples and encapsulated them with epoxy such that only the basal plane site (side) of graphene was exposed to the solution [30]. Optical images of the graphene and prepared graphene electrodes are shown in Fig. 3.16.

As shown in Fig. 3.17 (depicting the current response for the case of mono-, bi- and multi-layer graphene), sigmoidal currents were observed using the potassium ferri-/ferro- cyanide redox probe due to the exposed graphene surface effectively being a large microelectrode [30]; Chap. 2 overviews how changes in the electrode geometry can give rise to different mass transport and hence different voltammetric signatures.

Given that the edge of the graphene is covered with insulating epoxy and only its basal plane sites are exposed, it is surprising to observe any voltammetry at all. The reason for this observed voltammetry is that defects across the graphene surface reside [30], where there is a missing lattice atom and as such a dangling bond is exposed providing electrochemically reactive sites. Figure 3.18 shows a typical defect observed in graphene with TEM, Density Functional Theory (DFT) simulation of a graphene defect and an experimentally observed defect via STM [31]. Note also that the effect of defects on HOPG is well known, in that a 1 % defect density is estimated to result in a 10^3 factor increase in the heterogeneous electron transfer rate constant [32]. Dryfe et al. demonstrate that while their graphene surface has a low level of defects, fast electron transfer is observed due to the defects that are present on the graphene surface [30] and resultantly a similar voltammetric response is observed at bi-layer graphene (see Fig. 3.17) due to the top graphene layer only being exposed. Such work indicates that surface defects are extremely important in obtaining fast electron transfer rates, which has been shown for pristine graphene (see later) [33].

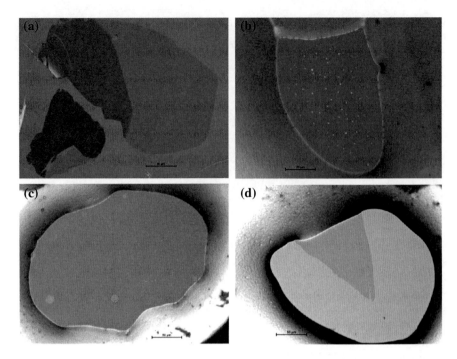

Fig. 3.16 Optical micrographs of monolayer graphene samples. Sample 1 is shown *before* (**a**) and *after* (**b**) masking; in image (**b**) edges are completely masked. Sample 2 is shown to contain holes in panel (**c**); in panel (**d**) the exposed part of sample 3 is *triangular*, hence edges are exposed to the solution. Details of each sample can be found in the legend of Fig. 3.17. *Scale bars*; (**a, c** and **d**) 50 μm and (**c**) 20 μm. Reprinted with permission from Ref. [30]. Copyright 2011 American Chemical Society

In the work of Dryfe [30], defects across the surface of the graphene are most probably due to the mechanical stresses involved in obtaining graphene from graphite using the 'scotch tape' (mechanical exfoliation) method. The sigmoidal response is likely due to the small size of the graphene sheet acting like a microelectrode (see Fig. 3.17) which is complicated further if the graphene is recessed by the epoxy. Note that defects across the basal surface of graphene are hard to determine and one approach is to use TEM and Scanning Tunnelling Microscopy (STM). Figure 3.18 shows a TEM image of a defect in graphene along with simulated structures and a STM image of a single vacancy; clearly determining defects is a challenging task.

The above reports are currently, at the time of writing this *Handbook*, the only two examples in the literature where individual graphene crystals have been electrochemically probed on the micro-scale and the reason as to why this is, is due to the large amount of effort that one has to undertake in order to perform such experiments. Clearly these are fundamental studies with the fabrication not scalable such that the most common approach to utilise graphene is to immobilise it

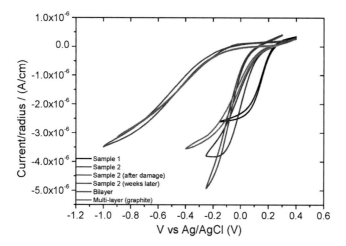

Fig. 3.17 Ferricyanide voltammetry: Current (normalised to electrode radius) versus potential response for the graphene monolayer samples (Samples 1 and 2: 1, monolayer contained no visible defects and its edges were completely masked—note however that although special attention was paid during the masking and preparation of samples in order to expose areas with the minimum number of defects, the authors acknowledge that to date it has not been possible to achieve a perfect, edge-free region; 2, monolayer contains several holes of ~ 10 µm diameter, hence some edge sites must be in contact with the electrolyte), a bilayer sample and the multilayer sample. Scan rate $= 5$ mVs^{-1}; concentration $= 1$ mM ferricyanide in 1 M KCl. Reprinted with permission from Ref. [30]. Copyright 2011 American Chemical Society

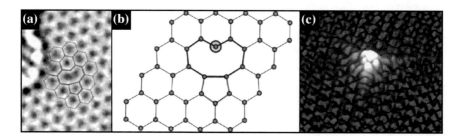

Fig. 3.18 a TEM image of a defect in a graphene lattice (Reprinted with permission from Ref. [34]. Copyright 2008 American Chemical Society); **b** Simulated atomic structure obtained via DFT calculations (Reprinted with permission from Ref. [31]. Copyright 2010 American Chemical Society); **c** An experimental STM image of a single vacancy, appearing as a protrusion due to an increase in the local DOS at the dangling bond (marked with a circle in *panel b*) (Reprinted with permission from Ref. [35]. Copyright 2010 The American Physical Society)

upon a suitable electrode surface such that one is effectively averaging a response over that of the graphene domains.

Key insights into the electrochemical reactivity of pristine graphene have been provided [33] through the modification of edge plane- and basal plane- pyrolytic graphite (EPPG and BPPG respectively) electrodes with pristine graphene, as is

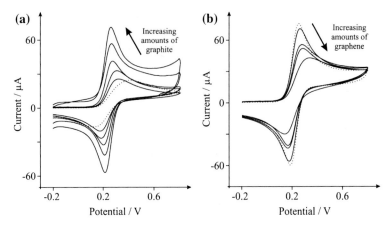

Fig. 3.19 Cyclic voltammetric profiles recorded utilising 1 mM potassium ferrocyanide(II) in 1 M KCl. **a** obtained using a BPPG electrode (*dotted line*) with the addition of increasing amounts of 2, 4, 50, 100, and 200 μg graphite (*solid lines*). **b** obtained using an EPPG electrode (*dotted line*) with the addition of increasing amounts of 10, 20, 30, and 40 ng graphene (*solid lines*). Scan rate: 100 mVs^{-1} (*vs.* SCE). Reproduced from Ref. [33] with permission from The Royal Society of Chemistry

common place in the literature in order to 'electrically wire' and connect to the graphene under investigation. The authors utilised a plethora of electroactive probes that have been commonly employed within the field (are well characterised on graphitic materials) and consequently have been well understood over many decades. Surprisingly, such work was the first to show a deviation from the wealth of literature on graphene at that time, in that pristine graphene was shown not to be as beneficial as previously reported [33].

Due to the type of graphene utilised, that is, high quality graphene with a low density of defects across the basal plane surface of the graphene sheet as well as a low oxygen content, it was observed that graphene exhibits slow electron transfer kinetics and as a result the electrochemical response (of the supported graphene layer) was found to actually block the underlying electrode surface [33]. The observed voltammetric response is presented in Fig. 3.19 for the electrochemical oxidation of ferrocyanide. Shown in Fig. 3.19a is the response of a BPPG electrode following modification with graphite, where it is well known that graphite has a large proportion of edge plane like- sites/defects and hence when one immobilises the graphite, the underlying electrode surface which exhibited slow electron transfer (note the large peak-to-peak separation, ΔE_P) exhibits a change (improvement) in the voltammetric signature due to the surface becoming populated with edge plane sites. In the case of graphene however (Fig. 3.19b), the reverse is observed, that is the introduction of graphene appears to be blocking the underlying electrode surface. In this case, when graphene was introduced onto a surface that exhibited fast electron transfer rates and a high degree of edge plane sites (an EPPG electrode), the immobilised graphene blocked the fast electron transfer occurring at the underlying surface, reducing the overall electrochemical activity, which can be attributed via the fundamental knowledge on graphite electrodes (see Sect. 3.1.2) to be due to the high proportion of relatively inert basal plane surface on pristine graphene as opposed to a small structural contribution from edge plane sites/defects [33].

In terms of the coverage of graphene over a supporting substrate, a key experimental parameter that needs to be considered, it was indicated that two "working zones" will arise when researchers immobilise graphene upon electrode substrates [33]. The first zone, '*Zone I*' corresponds to the modification of the electrode surface resulting in single- and few-layer graphene modified electrodes, which block the electrochemical response observed at the underlying electrode. Upon increasing amounts of graphene, the underlying electrode is continuing to be blocked (as shown in Fig. 3.19b). This is since the material (graphene) that is being immobilised has a low proportion of edge plane sites since the proportion of edge plane sites to basal plane sites (within its geometric structure) are extremely low and given its pristine nature, edge plane sites/defects across the basal surface are negligible. Upon the addition of more graphene, a '*Zone II*' becomes evident [33]. This is where several/ significant layers of graphene are observed (viz *quasi*-graphene [36] and graphite) which leads to an increment in the density of edge plane sites (due to its geometric structure) and thus improved voltammetry via increased heterogeneous electron transfer rates (as is evident in Fig. 3.19a). This response continues until a limit is observed, typically from the instability of the graphene upon the underlying electrode surface/support. Clearly the coverage of graphene is a key parameter in graphene electrochemistry, where the incorrect use/characterisation of a graphene modified surface could mislead those that are actually observing graphite (but believe they are using graphene) into misreporting the benefits of graphene, i.e. if working in *Zone II*. Note that recently a '*Zone III*' has been shown to exist in that, when excessive amounts of graphene is immobilised thin-layer effects dominate, giving the false impression of electro-catalysis; Chap. 2 overviews this concept.

Figure 3.20 highlights the change in the structure of the electrode surface from introducing graphene and the resultant electrochemical responses expected. Figure 3.20a shows a cyclic voltammetric profile as typically observed at an edge plane HOPG electrode assumed to possess fast electron transfer kinetics and following the immobilisation of single-layer graphene (Fig. 3.20b) where an incomplete coverage of the surface is realised. Effectively one is replacing a highly efficient and electrochemically reactive surface with graphene which has a low proportion of edge plane sites and no defects across the basal plane surface of the graphene, giving rise to the observed voltammetry with an increased ΔE_P indicating a departure towards slower electron transfer kinetics [33]. Following complete single-layer coverage (Fig. 3.20c) of graphene the ΔE_P increases which is firmly in *Zone I* as identified above. As more graphene is immobilised (Fig. 3.20d), a departure from single-layer, or approximate single layer/double and few layer (*quasi*-graphene) [36] is evident to that of multi-layer graphene (viz graphite) where one is now in *Zone II*, such that the voltammetric response heads

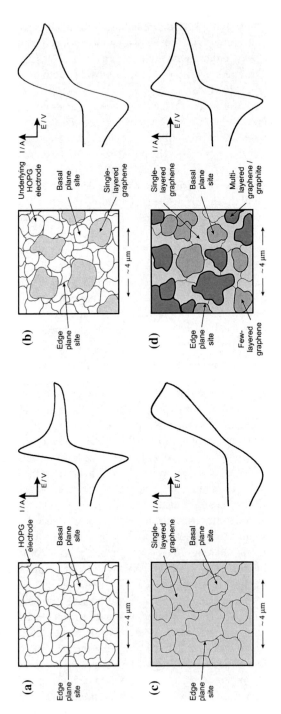

Fig. 3.20 Schematic representation of the effect on the cyclic voltammetric profiles that will be observed for a *HOPG* electrode following modification with differing coverages of graphene using a simple outer-sphere electron transfer redox probe. **a** Represents an unmodified *HOPG* electrode surface where fast electron transfer kinetics are observable **b** after modification with graphene leading to incomplete coverage where reduced electron transfer rates occur **c** after modification with graphene leading to complete single layer coverage where due to the large basal content of graphene (with little edge plane contribution) poor electrochemical activity is observed where electron transfer is effectively blocked, and **d** after continual modification with graphene leading to layered structures with increased edge plane sites available (origin of fast electron transfer) and thus an improvement in the electrochemical response is observed. Reproduced from Ref. [33] with permission from The Royal Society of Chemistry

back towards that originally observed for HOPG (Fig. 3.20a) due to the now large proportion of edge plane sites upon the electrode surface [33].

Thus, Brownson et al. has shown that given the geometric structure of graphite (multiple layers of stacked graphene), by its very nature it possesses a larger proportion of edge plane sites than that of single layer graphene and thus the former exhibits improved electrochemical activity, heterogeneous electron transfer kinetics, over that of the latter [33].

Returning to the case of immobilising pristine graphene onto a HOPG surface for electrochemical investigation (see above), insights from Brownson et al. [33] reveal that the underlying (supporting) electrode surface plays an important role, as does the orientation of the immobilised graphene. SEM images in the above work revealed that coalesced graphene 'folds' over edge plane sites of the underlying electrode, potentially explaining the blocking effect observed when graphene is introduced. Upon further additions of graphene, orientation with the edge plane sites of the underlying electrode results in a vertically aligned or disorder graphene surface and hence a beneficial increase in the electrochemical response is observed due to the increment in the proportion of edge plane sites accessible for electron transfer [33]. In this model it is assumed that the immobilised graphene adopts a similar architecture to that of the underlying electrode since the graphene has a distributed electron density of the planar–basal site (π–π) which will be disturbed by the high electron density of the underlying edge sites of the graphene (the EPPG) such that it effectively 'aligns' with the underlying electrode surface as this arrangement reflects the lowest energy settlement [33]. Due to the high number of graphene sheets on the EPPG electrode the graphene sheets will stack (as a continuation of the edge planes) in parallel to each other in order to fit the limited space of the EPPG surface. In the case of graphene upon a BPPG electrode surface the graphene will follow the same architecture as presented by the BPPG sheets, meaning that the graphene will stack planar on the BPPG due to π–π stacking: Fig. 3.21 depicts SEM images showing this concept.

Last, insights from DFT simulations on different sizes of graphene reveal a greater electron density at the edge of pristine graphene which confirms the observations by Brownson et al. [33] and in other work [37, 38] such that, similar to that observed for HOPG, the peripheral edge of graphene as opposed to its side acts electrochemically akin to that of edge plane sites and the latter to that of basal plane sites; in this case pristine graphene assumes no defects (defect sites, missing atoms, dangling bonds etc.) across the surface of the graphene, the introduction of which will beneficially contribute to the electrochemical activity of graphene. Note that in the case of graphite a greater density of edge plane sites is well known to result in an improved electrochemical reactivity [1, 4–6]; but until the above reports this was lesser reported for graphene.

In support of the above work it has been shown, using SECM to study correlations in monolayer and multilayer graphene electrodes grown via CVD, that in terms of these layered structures, single layer graphene exhibits the lowest electrochemical activity and that activity increases systematically when increasing the number of graphene layers, to a situation where the flakes are so active that the

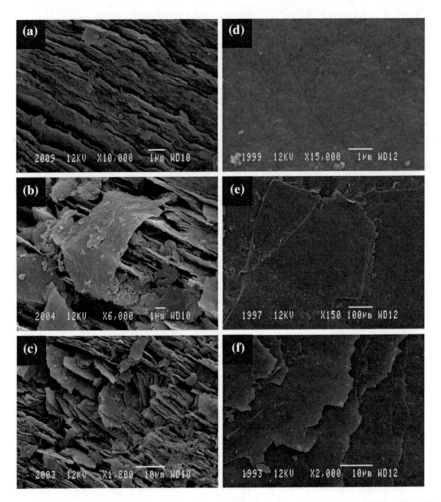

Fig. 3.21 SEM images of an unpolished EPPG electrode before (**a**) and after modification with low (**b**) and high (**c**) coverage's of graphene, and additionally an unpolished BPPG electrode before (**d**) and after modification with low (**e**) and high (**f**) coverage's of graphene. Reproduced from Ref. [33] with permission from The Royal Society of Chemistry

electron transfer process becomes nearly electrochemically reversible at greater than 7 layers (viz graphite) [39]. Figure 3.22 shows the electrochemical current as a function of increasing graphene layers using SECM which shows that single layer graphene exhibits the lowest electrochemical activity [39]. Such work confirms the work of Brownson et al., that states 'single layer (and bi- and few- layer (*quasi*-graphene)) graphene is not such a beneficial electrode material (when contrasted to graphite in terms of the heterogeneous electron transfer kinetics) and that in fact, in cases where beneficial electron kinetics are reported at graphene, it is likely that researchers are putting large quantities of graphene upon their electrode surfaces such that there is a large deviation from graphene to that of

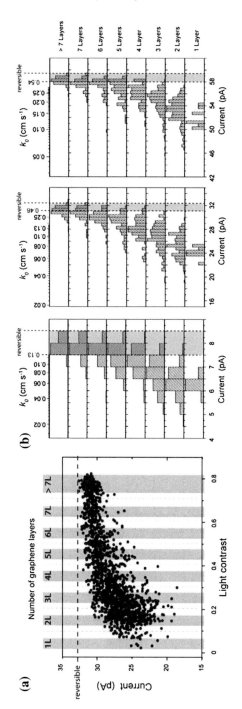

Fig. 3.22 a Pixel-by-pixel correlation between the electrochemical current map and the number of graphene layers. **b** Histograms of the electrochemical current and standard rate constant, k^0, for each defined number of CVD graphene layers, for potentials E_1, E_2, and E_3 (from left to right). The electrochemical oxidation of 2 mM FcTMA$^+$ (30 mM KCl) is used as the voltammetric probe with E_1, E_2, and E_3 corresponding to points on the voltammetric response of the FcTMA$^+$. The *dashed line* in (**a**) and the *blue area* in (**b**) denote the conditions where the electron transfer process becomes entirely reversible. Reprinted with permission from Ref. [39]. Copyright 2012 American Chemical Society

graphite and thus false claims of improvements in the electrochemical processes of graphene are generally reported (which should instead be attributed to graphite and/or other graphitic structures).

Other notable work adding to the fundamental understanding of graphene is by Lim et al. [40], who investigated the effect of edge plane defects on the heterogeneous charge transfer kinetics and capacitive noise at the basal plane of CVD fabricated epitaxial graphene (prepared on a silicon carbide substrate) using inner-sphere and outer-sphere redox mediators. The authors showed that the basal plane surface of graphene exhibits slow heterogeneous electron transfer kinetics, interestingly however, when electrochemically anodised (increasing the degree of oxygen-related edge plane defects) they found that the defects created on its surface resulted in the improvements of election transfer rates that surpass those observed for pristine graphene, Glassy Carbon (GC) and Boron Doped Diamond (BDD) electrodes [40]. Again, this work confirms the essential need for edge plane like-sites/defects on the surface of graphene (for improved electrochemical reactivity). Note, it is well known that the presence of oxygen related species on a carbon based electrode material can dramatically influence the observed electrochemical reactivity, either beneficially or detrimentally depending on the target analyte [2, 41–45]. Thus it could be inferred that the oxygen-related species purposely introduced onto the graphene surface in this case contribute to a hidden origin of the improved rate kinetics. However, this is not the case and the contribution from the oxygenated species residing on the graphene can be neglected since the authors utilised a range of electro-active species to study their graphene, ranging from simple outer-sphere electron transfer probes to surface sensitive inner-sphere species (see Box 3.1), and the observed trend was similar for all compounds.

BOX 3.1: Surface Sensitivity at Inner- and Outer- Sphere Redox Probes

A common approach within electrochemical studies in order to greater understand the material under investigation is the utilisation of inner-sphere and outer-sphere redox mediators/probes. Such electron transfer processes differ significantly according to the 'sensitivity' of their electron transfer kinetics to the surface chemistry of the carbon electrode/material under investigation in terms of the surface structure/cleanliness (defects, impurities or adsorption sites) and the absence/presence of specific oxygen containing functionalities, that is, variations in k^o with the condition of the electrode surface [1, 46].

Outer-sphere redox mediators (see Fig. 3.23 for examples) are termed *surface insensitive* such that k^o is not influenced by the surface oxygen-carbon ratio, surface state/cleanliness in terms of a surface coating of a monolayer film of uncharged adsorbates, or specific adsorption to surface groups/sites [1]. There is no chemical interaction or catalytic mechanism involving interaction (i.e. an adsorption step)

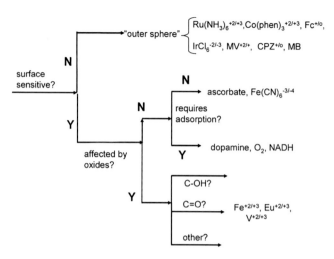

Fig. 3.23 Classification of redox systems according to their kinetic sensitivity to particular surface modifications on carbon electrodes. Reprinted with permission from Ref. [47]. Copyright 1995 American Chemical Society

with the surface or a surface group—such systems often have low reorganisation energies [1, 46]; in this case the electrode merely serves as a source (or sink) of electrons and as such outer-sphere systems are sensitive primarily to the electronic structure due to the electronic DOS of the electrode material [1, 46].

Inner-sphere redox mediators (see Fig. 3.23 for examples) are termed *surface sensitive* in that the k^o is strongly influenced by the state of the electrode surface (surface chemistry and microstructure) via specific electro-catalytic interactions that are inhibited significantly if the surface is obscured by adsorbates (or impurities). Such interactions can also depend strongly on the presence (or absence) of specific oxygenated species which give rise to either beneficial or detrimental effects [1, 46]. In this case systems are more largely affected by surface state/structure and/or require a specific surface interaction, being catalysed (or inhibited) by specific interactions with surface functional groups (adsorption sites) rather than the DOS as such systems generally have high reorganisation energies [1, 46].

The observation of differing responses when using varied inner- and outer- sphere redox probes allows insights to be deduced regarding the state of the surface structure of the electrode material in question. McCreery [62, 87] has provided a "road map" for commonly utilised redox probes, as shown in Fig. 3.23, which allows researchers to clarify from experimental observations the redox systems and how they are affected.

Other notable work that supports the findings that edge plane sites of graphene are dominantly reactive over that of its basal plane has been published by Keeley et al. [48, 49], in which the authors sonicated graphite powder in dimethylformamide for 72 h in order to achieve exfoliation of graphene nanosheets whilst precluding the need for chemical oxidation and as such alleviating any unnecessary contributions from the presence of surface oxygenated species. TEM analysis showed that 90 % of the resultant nanosheets contained five or fewer graphene layers and the lateral dimensions were mostly less than 1 μm, leading to a much higher density of edge plane- like sites than the parent graphite, as was confirmed by Raman spectroscopy. Cyclic voltammetric measurements with common redox probes confirmed that nanosheet-functionalised electrodes had larger active areas and exhibited far more rapid electron transfer than the plain/unmodified electrodes. Due to the absence of oxygen-containing groups in the solvent-exfoliated graphene, the observed electrochemical activity was attributed to originate from the numerous edge plane- like sites and defects on the graphene nanosheets [48, 49]. Moreover, other work has considered the electrochemical activities of open and folded graphene edges (in which the folded edges were structurally more similar to basal plane) where it was demonstrated that the heterogeneous electron transfer rate is significantly lower on folded graphene edges than that of open graphene edges (as is evident in Fig. 3.24 where larger ΔE_P is observed at the former over that of the latter) [50]; again such work concurs with the concept that the edge plane of graphene is the origin of electron transfer [33]. It is clear that there is a requirement to have exposed edges of graphene to achieve optimal electrochemical activity (fast electron transfer rates), or equivalently to have a high density of edge plane like- sites/defects across the graphene surface.

As highlighted in Chap. 1 different preparative methods of graphene result in structures that have greatly varied densities of edge plane defects [12] and thus it has been shown that the method of graphene preparation consequently has a dramatic influence on the materials properties and electrochemical reactivity [12, 51]. Furthermore, surface defects can be selectively introduced into the graphene structure post-synthesis, for example through the use of ion or electron irradiation, selective oxidation (with optional reduction), or by mechanical damage [31]. Note that incorporation of dopants/foreign atoms (i.e. nitrogen doping, see later) or the introduction of functionalities (i.e. oxygenated species) in addition to the formation of composite (or novel three-dimensional) [36] graphene based materials have also all been reported to alter the electrochemical properties of graphene either beneficially or detrimentally, that is, in terms of the observed heterogeneous electron transfer rates, DOS, intrinsic catalytic attributes and influences on surface adsorption/desorption processes (see Chap. 4 for examples of how this can be utilised to result in beneficial performances being observed at modified graphenes for various applications) [20, 41–45, 52, 53]. Importantly, if controllable and reproducible defect densities of graphene can be achieved and quantified [38], as has been shown to be the case at CVD graphene [12], then the electrochemical reactivity of graphene can be optimised and efficiently tailored when designing

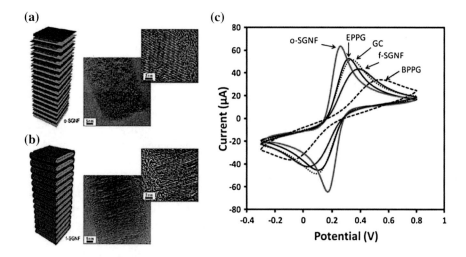

Fig. 3.24 Electrochemistry at folded edges of graphene sheets. Schematic drawing and TEM micrographs of **a** open and **b** folded graphene edge nanostructures. The drawing is not to scale and it should be noted that the "opening" of the folded edge nanostructure is to illustrate the inner structure of the fibre. The structural difference between open (**a**) and folded (**b**) edges are clearly visible in the detailed TEM images on the right (Note, scale bars represent 5 and 2 nm in each case). (**c**) Cyclic voltammograms of 5 mM potassium ferrocyanide(II) in 0.1 M KCl supporting electrolyte on open graphene structure (*green*; *o-SGNF*), folded graphene structure (*red*; *f-SGNF*), EPPG (*solid line*), GC (*dotted line*) and BPPG (*dashed line*) electrodes. Scan rate 100 mVs^{-1} (*vs.* Ag/AgCl reference electrode). Reproduced from Ref. [50] with permission from The Royal Society of Chemistry

graphene based devices with dedicated properties to achieve new functions/applications (to exhibit either fast or slow heterogeneous electron transfer—or to possess specific binding/attachment sites in the case of functionalisation). Thus effectively graphene can provide an electrochemically beneficial platform where it is possible to modify the graphene structure so that the properties of this material suit specific needs; such tailoring and versatility is of eminent importance for practical applications as well as for academic research.

3.2.2 Effect of Surfactants on the Electrochemistry of Graphene

As highlighted in Sect. 1.2 suspensions of graphene in a liquid are often (but not always) stabilised by surfactants, which are routinely incorporated into the fabrication of commercially available graphene to reduce the likelihood of the graphene sheets coalescing. When this is the case a tentative approach must be employed when utilising such graphene solutions within electrochemistry [54–57]. It has been established that some surfactants [58], for example sodium cholate [54–57], exhibit measurable electrochemical activity and can thus contribute towards or even dominate the electrochemical properties and performance of the stabilised graphene, such that highly negative effects on the interpretation of data have been observed [54–57]. This was demonstrated to be the case towards the detection of β-nicotinamide adenine dinucleotide (NADH) and acetaminophen (APAP; paracetamol) as well as in the stripping voltammetry of heavy metals [54, 55]. These interferences/effects also extended to energy storage applications where it was demonstrated that surfactants themselves provide a higher capacitance than graphene, thus one must be cautious when attributing beneficial effects to graphene in these instances [57].

This work poses a highly important warning to the graphene community to always consider the effect/influence of any surfactant/solvent that is used to aid graphene dispersion. It is clear that appropriate control experiments need to be employed before the beneficial electrochemistry of graphene can be correctly reported and such control measures are thus required in future experiments in order to sufficiently de-convolute the true performance of graphene. Such a warning can be extended to other aspects of graphene electrochemistry, including the requirement for appropriate control and comparison experiments when determining the contribution to the electrochemical response of graphene in terms of the presence of graphitic impurities and oxygenated species (see Chap. 4).

3.2.3 Metallic and Carbonaceous Impurities on the Electrochemistry of Graphene

When carbon nanotubes (CNTs, which are effectively rolled up graphene structures) were first utilised in electrochemical applications they were found to contain metallic impurities as a result of their CVD fabrication, which contributed or dominated their observed electrochemical response [59, 60]. Similarly, it has been shown that graphene fabricated from graphite, via chemical oxidation of natural graphite followed with thermal exfoliation/reduction, can contain cobalt, copper, iron, molybdenum and nickel oxide particles which can influence the electrochemistry of graphene towards specific analytes and has potential to lead to inaccurate claims of the electro-catalytic effect of graphene [61, 62].

Note that such impurities (as noted above) can be avoided since, in the methodology, the grade of graphite purchased will largely dictate the final product and the highest purity graphite should be used (as well as high purity aqueous and non-aqueous solvents) to alleviate such problems. Nonetheless, as with Sects. 3.2.2 and 3.2.6 this valuable work highlights the importance of sufficient control experimentation when exploring graphene within electrochemistry. Note also however, that in certain cases the presence of metallic impurities may be beneficial with respect to the observed electrochemical response and thus indeed purposely incorporated as part of the fabrication process with the aim of producing noble metal (Pd, Ru, Rh, Pt, Au or Ag) doped/decorated graphene hybrid structures that exhibit beneficial electronic properties, for use in electrocatalytic sensing for example [63]; nevertheless, in such cases it is crucial that the quantity and quality of the said metals is cautiously analysed/controlled and reported (along with the appropriate control experimentation).

Last, the effect of carbonaceous debris (in addition to edge plane defects as covered at the end of Sect. 3.2.1) has been explored upon the electrochemical behaviour of reduced graphene oxide (GO), and was reported to strongly affect the observed electrochemical response [64]. It is important to take from this work that when GO is reduced to graphene, pristine graphene does not result, rather a graphene that possesses strongly adhered carbonaceous debris and edge plane defect sites which are created during the synthesis of GO [64]. Other comparable work has also shown this to be the case, where the carbonaceous debris significantly impacted the observed electrochemical response (beneficial in this case for electroanalysis, as is expected given the insights gained in Sect. 3.2.1) [64]. Interesting work has been performed by Tan and co-workers [65] utilising SECM to study the reactivity of surface imperfections present (i.e. edge plane defects or carbonaceous debris) on monolayer graphene, revealing that specific sites across the surface of monolayer graphene that have a large concentration of defects (introduced either through deliberate mechanical damage or through chemical oxidation (as with reduced GO)) are approximately 1 order of magnitude more reactive, compared to more pristine graphene surfaces, toward electrochemical reactions [65]. Of further importance, the authors were able to successfully passivate the activity of graphene defects by carefully controlling the electro-polymerisation of o-phenylenediamine so that a thin film of the polymer was formed (which was found to be insulating in nature toward heterogeneous electron transfer processes): thus it was demonstrated that SECM can be utilised for detecting the presence of (and "healing") surface defects on graphene; providing a strategy for in situ characterisation and control of this fascinating material and enabling optimisation of its properties for select applications as stated in Sect. 3.2.1 [65].

3.2.4 Electrochemical Reports of Modified (N-doped) Graphene

As mentioned earlier, through the modification of graphene one can tailor its properties to produce *task specific graphene*. Although there are a wide range of modified graphenes that can be synthesised (the electrochemical applications of which are fully explored in Chap. 4), in this section we focus solely on the use of nitrogen doped (*N*-doped) graphene to give readers a general overview of how the resultant properties of the modified graphene differ from that of pristine graphene and hence can lead to beneficial outcomes.

The chemical doping of graphene falls into two key areas: (i) the adsorption of organic, metallic and gaseous molecules/compounds onto the graphene surface; and (ii) substitutional doping, where heteroatoms are introduced into the graphene lattice, such as nitrogen and boron [66–69]. Both approaches have been reported to alter the electronic properties of graphene (including the DOS, which in turn influences the heterogeneous electron transfer properties observed as was highlighted in Chap. 2) [66, 70], where for example doping with boron or nitrogen atoms allows graphene transformation into either a p- or n-type semiconductor respectively [70–73]. Generally the doping of graphene with nitrogen is favoured and widely pursued because it is (a nitrogen atom) of comparable atomic size and the fact that it contains five valence electrons that are available to form strong valence bonds with carbon atoms [74].

Wang and co-workers provide a thorough overview of the synthesis of N-doped graphene [66], in which the authors overview the many fabrication routes available, which include direct synthesis (such as through CVD), solvothermal approaches, arc-discharge, and post-synthesis treatments such as thermal, plasma and chemical methods [66, 75]; Table 3.1 overviews these various methodologies.

Through the doping of graphene to produce N-doped graphene, three common bonding configurations arise, which are: quaternary N (or graphitic N), pyridinic N, and pyrrolic N. As shown in Figs. 3.25 and 3.26, pyridinic N bonds with two C atoms at the edges or defects of graphene contribute one p electron to the π system while Pyrrolic N refers to N atoms that contribute two p electrons to the π system, although unnecessarily bond into the five-membered ring, as in pyrrole. Quaternary N refers to N atoms that substitute for C atoms in the hexagonal ring. Among these nitrogen types, pyridinic N and quaternary N are sp^2 hybridised and pyrrolic N is sp^3 hybridised. Apart from these three common nitrogen types, N oxides of pyridinic N have been observed in both the N-graphene and N-CNT studies where the nitrogen atom bonds with two carbon atoms and one oxygen atom [66, 96, 97].

In terms of the electrochemical application of doped graphene, one area where N-doped graphene is being extensively studied is as a replacement for platinum in fuel cells [68, 77, 91, 98–101]. Other work has reported the electrochemical oxidation of methanol (again for fuel cell applications), fast electron transfer kinetics for glucose oxidase, and high sensitivity and selectivity for glucose bio-sensing, in addition to the direct sensing of hydrogen peroxide [68, 77, 91, 98–101].

3.2.5 The Electrochemical Response of Graphene Oxide

GO is by no means a new material, with first reports emerging around 1859. Structurally it constitutes single atomic layers of functionalised (oxygenated) graphene that can readily extend up to tens of μm in lateral dimension. GO can be viewed as an unconventional type of soft material as it carries the characteristics of polymers, colloids, membranes and is an amphiphile [102, 103]. The specific structure of GO is debatable and Fig. 1.4 (found in Chap. 1) shows the various

Table 3.1 Nitrogen-doping methods and nitrogen (N) concentration on graphene. Reprinted with permission from Ref. [66]. Copyright 2012 American Chemical Society

No.	Synthesis method	Precursors	N content, at. %	Application/reference
1	CVD	Cu film on Si substrate as catalyst, CH_4/NH_3	1.2–8.9	FET / [70]
2	CVD	Cu foil as catalyst, NH_3/He	1.6–16	ORR / [76]
3	CVD	Ni film on SiO_2/Si substrate as catalyst, $NH_3/CH_4/H_2/Ar$ (10:50:65:200)	4	ORR / [77]
4	CVD	Cu foil as catalyst, acetonitrile	~9	Lithium battery / [78]
5	CVD	Cu foil as catalyst, pyridine	~2.4	FET / [79]
6	Segregation growth	Carbon-contained Ni layer on nitrogen-contained boron layer	0.3–2.9	FET / [80]
7	Solvothermal	Li_3N/CCl_4 (NG1) or $N_3C_3Cl_3/Li_3N/CCl_4$ (NG2)	4.5 (NG1) or 16.4 (NG2)	ORR / [81]
8	Arc discharge	Graphite/H_2He/pyridine (NG1) graphite/H_2/He/NH_3 (NG2) transformation of nanodiamond/He/pyridine (NG3)	0.6 (NG1), 1 (NG2), 1.4 (NG3)	[82, 83]
9	Thermal treatment	N^+ ion-irradiated graphene, NH_3	1.1	FET / [84]
10	Thermal treatment	Graphite oxide after thermal expansion, NH_3/Ar	2.0–2.8	ORR / [85]
11	Thermal treatment	GNR, NH_3	NSD	FET / [86]
12	Thermal treatment	GO, NH_3/Ar (10 % NH_3)	~3–5	FET / [87]
13	Thermal treatment	GO, NH_3	6.7–10.78	Methanol oxidation / [88]
14	Thermal treatment	GO, melamine	7.1–10.1	ORR / [89]
15	Plasma treatment	Graphite oxide after thermal expansion, N_2 plasma	8.5	ORR / [68]
16	Plasma treatment	Graphite oxide after thermal expansion, N_2 plasma	3	ORR / [90]
17	Plasma treatment	Chemically synthesised graphene, N_2 plasma	~1.3	Biosensors / [91]
18	Plasma treatment	GO, treat with H_2 plasma first, then treat with N_2 plasma	1.68–2.51	Ultracapacitor / [92]
19	Plasma treatment	Mechanically exfoliated graphene or bilayer graphene grown by CVD, NH_3 plasma	NSD	FET / [93]
20	N_2H_4 treatment	GO, N_2H_4, NH_3	4.01–5.21	[94]
21	N_2H_4 treatment	Graphite oxide after thermal expansion, N_2H_4	1.04	Electrochemical sensor / [95]

Abbreviations: *CVD* chemical vapour deposition; *GNR* graphene nanoribbons; *GO* graphene oxide; *NSD* not stated; *FET* field-effect transistor; *ORR* oxygen reduction reaction

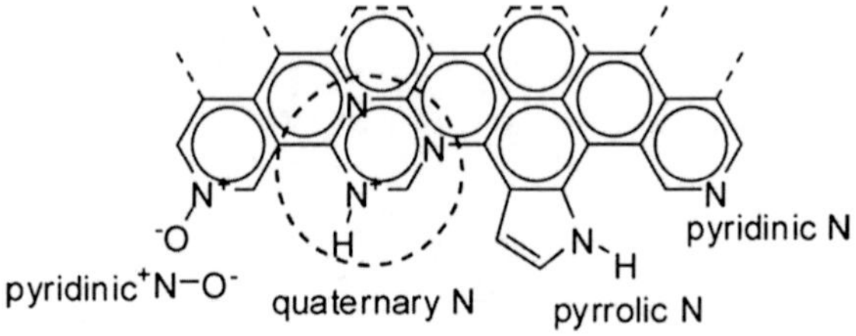

Fig. 3.25 Bonding
configurations for nitrogen
atoms in *N*-graphene.
Reprinted with permission
from Ref. [66]. Copyright
2012 American Chemical
Society

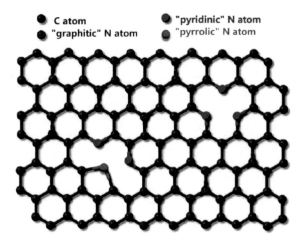

pyridinic$^+$N$-$O$^-$ quaternary N   pyrrolic N   pyridinic N

Fig. 3.26 Schematic
representation of *N*-doped
graphene. The *blue, red,
green,* and *yellow spheres*
represent the C, "graphitic"
N, "pyridinic" N, and
"pyrrolic" N atoms in the *N*-
doped graphene, respectively.
Evidence of the N-doped
graphene structure is
supported by XPS. Reprinted
with permission from
Ref. [70]. Copyright 2009
American Chemical Society

● C atom   ● "pyridinic" N atom
● "graphitic" N atom   ● "pyrrolic" N atom

models that can exist along with the most recent proposition shown in Fig. 3.27
which accounts for weaknesses in the models proposed in Fig. 1.4 [104].

As highlighted in Chap. 1 GO is useful, in that, a common approach to fabricate
graphene is to chemically, thermally or electrochemically reduce GO. Of note is
work by Zhou et al. [105] who have reported the electrochemical reduction of GO
films, which has resulted in a material that has a low O/C ratio. Such reduction
proceeds via the following electrochemical process (where ER implies electrochemically reduced):

$$GO + aH^+ + be^- \rightarrow ER:GO + cH_2O \qquad (3.7)$$

Of importance here is that the electrochemical reduction gives rise to voltammetric reduction waves at highly negative potentials. Figure 3.28 depicts the
electrochemical reduction of GO where the cathodic peak potentials are observed
to shift negatively as the pH increases [105]. This is believed to be a consequence
of the protonation involved in the electro-reduction process (see Eq. 3.7), which is
facilitated at lower pH values.

Exploring the electrochemistry of GO further, it is important to note that in
other work the cyclic voltammetric response of GO has been shown to be unique
towards specific electron transfer redox probes [106]. Given this fact, such unique

(a)

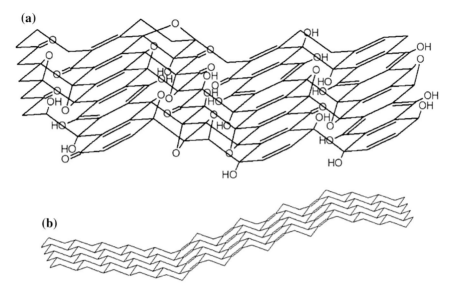

(b)

Fig. 3.27 A new structural model of GO: **a** surface species and **b** folded carbon skeleton. Reprinted with permission from Ref. [104]. Copyright 2006 American Chemical Society

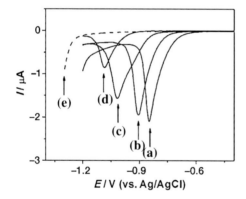

Fig. 3.28 Linear sweep voltammograms (in phosphate buffer solution (PBS)) of a GC electrode in contact with $\sim$7-mm-thick GO film supported on quartz (5×4 cm^2) at pH values of 4.12 (*a*), 7.22 (*b*), 10.26 (*c*) and 12.11 (*d*). Note that (*e*) represents a response recorded at the unmodified/bare GC electrode suspended in Na-PBS (1 M, pH 4.12). Reproduced from Ref. [105] with permission from Wiley

voltammetry can thus be used (as a characterisation technique) to ensure that GO has been fully transformed to graphene by exploring the voltammetric response before and after the chosen treatment has been applied [106]. Figure 3.29a shows the voltammetric response of increasing amounts of GO immobilised on an electrode surface (using the outer-sphere redox probe hexaammine-ruthenium(III) chloride), which is directly compared with that of increasing graphene additions

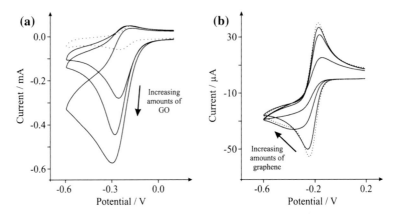

Fig. 3.29 Cyclic voltammetric profiles recorded towards 1 mM hexaammine-ruthenium(III) chloride in 1 M KCl. Scan rate: 100 mVs^{-1} (*vs.* SCE). **a** obtained using an EPPG electrode (*dotted line*) after modification with increasing depositions of 1.38, 2.75 and 8.25 µg GO (*solid lines*). Reproduced from Ref. [106] with permission from The Royal Society of Chemistry. **b** Obtained using an EPPG electrode (*dotted line*) with the addition of increasing amounts of 100, 200 and 300 ng graphene (*solid lines*). Reproduced from Ref. [33] with permission from The Royal Society of Chemistry

(Fig. 3.29b). It can be readily observed that unique voltammetry is evident at the GO modified electrode, which is quite different to that observed at the graphene modified electrode, where it is thought that in the case of GO the oxygenated species present contribute to a catalytic process, the *EC'* reaction (see Sect. 2.5), where a first 'electron transfer process' (E process), as described generally as [106]:

$$A + e^- \rightleftharpoons B \quad (E\ Step) \tag{3.8}$$

is then followed by a 'chemical process' (C process) involving the electro-generated product, *B*, which regenerates the starting reactant, *A*, as described by [106]:

$$B + C \xrightarrow{k} A + products \quad (C\ Step) \tag{3.9}$$

The voltammetric response arises as the amount of *C* is increased; which is attributed to the oxygenated species of the GO in this case [106]. Bear in mind that this response is unique to hexaammine-ruthenium(III) chloride and also occurs to a lesser extent for potassium hexachloroiridate(III) [106]. Crucially, the observed voltammetric reduction waves evident in Fig. 3.28 coupled with the voltammetry observed in Fig. 3.29 can be used as a measure to determine whether GO has been efficiently (electrochemically) reduced prior to its application in a plethora of areas [105, 106].

3.2.6 Electrochemical Characterisation of CVD Grown Graphene

The electrochemistry of CVD fabricated graphene can yield beneficial insights into the surface structure of the grown graphene layer [107, 108]. It has been demonstrated that through the careful choice of redox probes, different voltammetric responses can be observed, allowing insights into the structure and composition of the surface under investigation to be readily derived [107, 108].

In the case of CVD grown graphene that possesses a uniform (and complete) coverage of graphene, however with graphitic islands present randomly distributed across the surface; when using the outer-sphere redox probe 'hexaammine-ruthenium(III) chloride' to characterise the surface of the graphene electrode, a typical peak shaped cyclic voltammetric profile is observed and would indicate that a uniform 'graphene' film has been successfully fabricated (Fig. 3.30a). In this case, the outer-sphere electron transfer probe is surface insensitive (see Box 3.1) and thus the graphene electrode (namely *all* of the edge plane sites present) merely acts as a source (or sink) of electrons such that the majority of the geometric electrode area participates in the electrochemical reaction; resulting in heavily overlapping diffusional zones (see Fig. 3.30c, which is of course dependent on the applied scan rate) and hence a macro-electrode type response (cyclic peak shaped) is observed [107]. It is only by using redox probes with varying surface sensitivities however that material scientists are able to fully characterise their CVD grown graphene; we now consider the electrochemical response when utilising an inner-sphere redox probe.

If the redox probe is changed to an inner-sphere probe such as 'potassium ferrocyanide(II)', it is possible that a completely different voltammetric response is observed, as shown in Fig. 3.30b, where a steady-state voltammetric response is evident; this is usually seen at microelectrodes (see Chap. 2)—note that the voltammetric response in Fig. 3.30b is obtained after potential cycling. It is well documented that potassium ferrocyanide is highly surface sensitive (see Box 3.1) and exhibits a response dependent on the Carbon-Oxygen surface groups present on the surface of an electrode, ranging from beneficial to detrimental responses [41]. Since only a certain number of graphitic domains (that is, double-, few- and multi-layered graphene) residing on the surface are activated with the correct proportion of Carbon-Oxygen groups through potential cycling, only certain areas of the electrode surface become activated. This effectively produces a random array of active areas across the electrode surface, which have their own diffusion zones (see Fig. 3.30c), the majority of which are separated far from each other (under the applied voltammetric scan rate utilised in the given work) [107] such that these diffusion zones do not interact and thus sigmoidal voltammetry is observed. In this case, assuming that each activated zone/domain is akin to a microelectrode array, the electrochemical response, that is, a limiting current (see Chap. 2) is given by:

$$I_L = nFrCDN \qquad (3.10)$$

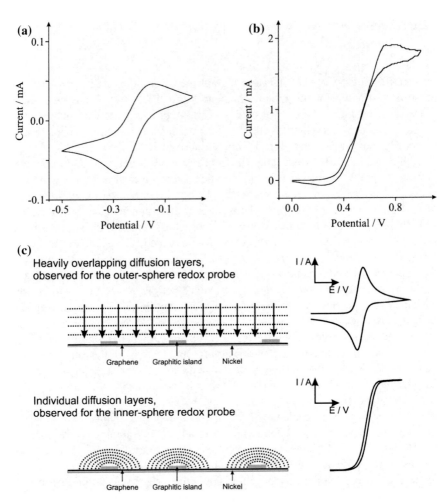

Fig. 3.30 **a** Cyclic voltammetric profiles recorded utilising 1 mM hexaammine-ruthenium(III) chloride in 1 M KCl, obtained using a CVD-graphene electrode (*dashed line*). **b** Cyclic voltammetric profiles recorded for 1 mM potassium ferrocyanide(II) in 1 M KCl using a CVD-graphene electrode. **c** A schematic representation of differing diffusion zones observable towards graphitic islands present upon CVD-graphene. **a** and **b** performed at a scan rate of 100 mVs^{-1} (*vs.* SCE). Figure reproduced from Ref. [107] with permission from The Royal Society of Chemistry

where n is the number of electrons, F the Faraday constant, C is the concentration of the analyte, D the diffusion coefficient of the analyte and r is the electrode radii. Note that N is the number of electrodes/activated zones/domains comprising the CVD surface and the magnitude of the voltammetry observed in Fig. 3.30b is simply "amplified" by N if the criteria that each diffusion zone is independent of each other and not overlapping, is apparent. It is apparent that the current is larger at the steady-state response (Fig. 3.30b) over that of the peak shaped response

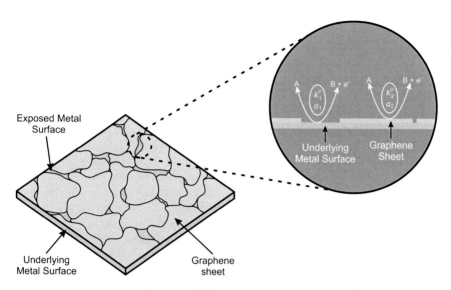

Fig. 3.31 A schematic representation of non-continuous CVD grown graphene electrode where the underlying metal surface is exposed to the solution, both with different Butler-Volmer characteristics. Note that the depending on the shape, size and reactivity of the given exposed surface (and that of the graphene), different voltammetry will be observed as per Chap. 2

(Fig. 3.30a) which is due to the above description, through being an array of microelectrdes/activated domains.

The importance of utilising different electrochemical redox probes in order to gain insights and effectively characterise potential electrode materials (viz Box 3.1) has thus been shown above. For instance if indeed a complete coverage of uniform single-layered pristine graphene had been present in the above case no voltammetric responses would have been observed. If defects occur across the basal plane surface then the voltammetry would appear very slow, slower than that observed for a BPPG with the response becoming more reversible (towards that of EPPG) as the global coverage of edge plane like- defects increased. Furthermore, other scenarios could exist, as explored below.

An interesting scenario can occur when material scientists grown their graphene via CVD and fail to perform adequate control experiments, as is common in the literature. That is, a beneficial electrochemical response is observed using the CVD grown graphene and control experiments using the underlying metal surface that the graphene has been grown upon, typically nickel [109, 110] or copper [11, 111] have been neglected to be performed using the same redox probe used to characterise the graphene surface. Figure 3.31 shows the case of a non-continuous graphene film where the underlying metal surface upon which the graphene was grown is exposed to the solution [107, 108]. In this case, we have a heterogeneous electrode surface which is reminiscent of the case of the HOPG surface as discussed in Sect. 3.1.2 where again, the two surfaces will have their own electrochemical activity (heterogeneous rate constant) towards the given redox probes;

that is, the graphene film, $k^0_{graphene}$ and the underlying metal surface $k^0_{metal\,surface}$ upon which the graphene has been grown.

In the first case, if the graphene film has a greater electrochemical activity over that of the underling metal surface, such that $k^0_{graphene} \gg k^0_{metal\,surface}$, the grown graphene domains will dominate the electrochemical response. Assuming that the coverage of graphene is of the macro-electrode scale this will giving rise to peak shaped voltammetry as the surface will be akin to a macroelectrode. However, note that as the graphene sheet becomes non-continuous and graphene islands are formed (which are randomly distributed across the underlying metal surface), each graphene domain will have its own diffusion zone and if these are not heavily overlapping (see Fig. 3.30c) such that these diffusion zones do not interact, sigmoidal voltammetry will dominate. Thus the above case will likely be observed as discussed in Fig. 3.30 with select redox probes: depending on the graphene's size, shape and orientation (as well as surface coverage) different voltammetry will be observed, as covered in Chap. 2. Note of course that this will also depend upon the chosen voltammetric scan rate, the diffusion coefficient of the electroactive target/analyte and the coverage of the dominant electrochemically active material.

In the second case of a non-continuous graphene film, the underlying metal surface (due to its reactivity) will act like an electrode and if the metal has favourable electrochemical properties towards the electrochemical analyte which is of interest, it will contribute to the observed electrochemical response or even dominate it; misleading researchers into thinking that graphene exhibits excellent electrochemical properties [107, 108]. In this case, assuming $k^0_{graphene} \ll k^0_{metal\,surface}$ the metal surface will dictate the electrochemical response depending on the coverage of the graphene film/islands. If the coverage of graphene is large, such that only nano- or micro- bands of the underlying metal are exposed, a microelectrode type response will be observed in the form of a steady-state response (sigmoidal voltammetry) where the magnitude of the current is amplified by the number of exposed domains and will be governed by Eq. (3.10). However, this will depend upon the distribution of the graphene islands and an alternative case might result in even macro-electrode type structures/areas of the underlying metal being exposed! This last case will obviously result in peak shaped voltammetry dominating, which after being originally theorised in Ref. [107], has been shown recently at ill formed graphene grown upon nickel and copper such that the exposed underlying electrode dominates over that of graphene (in this case due to the poor fabrication, the exposed surfaces are large irregular macro-sized domains) [112]. The key to determining if the underlying surface is contributing towards the observed electrochemical response is to run control experiments, that is, perform the same voltammetry using the underlying supporting material (viz no graphene) to determine the extent of the electrochemical activity.

Last, a really unique response might be observed, which was discussed in Sect. 3.1.2 with respect to HOPG electrode surfaces, such that two voltammetric peaks might be observed! In this case a first peak would arise from a material exhibiting fast electron transfer and a second peak from the graphene surface. The extent of the observed voltammetry will depend on the electrochemical activity of the two materials comprising the heterogeneous surface, the distance between active sites (diffusion zones), voltammetric scan rates, the diffusion coefficient of the electroactive target/analyte and the coverage of the dominant electrochemically active material.

It is evident, due to the above scenarios, that the electrochemistry of CVD grown graphene is highly fascinating. This section has also highlighted that materials scientists can beneficially utilise electrochemistry as a tool to assist in the characterisation of their fabricated 'graphene' material.

References

1. R.L. McCreery, Chem. Rev. **108**, 2646–2687 (2008)
2. D.A.C. Brownson, C.E. Banks, Analyst **135**, 2768–2778 (2010)
3. Web-Resource, http://www.nanoprobes.aist-nt.com/apps/HOPG%2020info.htm. Accessed 18 Jan 2013
4. T.J. Davies, M.E. Hyde, R.G. Compton, Angew. Chem. Int. Ed. **44**, 5121–5126 (2005)
5. T.J. Davies, R.R. Moore, C.E. Banks, R.G. Compton, J. Electroanal. Chem. **574**, 123–152 (2004)
6. C.E. Banks, R.G. Compton, Analyst **131**, 15–21 (2006)
7. Web-Resource, www.graphene-supermarket.com. Accessed 28 Feb 2012
8. A. Dato, V. Radmilovic, Z. Lee, J. Phillips, M. Frenklach, Nano Lett. **8**, 2012–2016 (2008)
9. M. Lotya, P.J. King, U. Khan, S. De, J.N. Coleman, ACS Nano **4**, 3155–3162 (2010)
10. Web-Resource, www.nanointegris.com. Accessed 28 Feb 2012
11. X. Li, C.W. Magnuson, A. Venugopal, R.M. Tromp, J.B. Hannon, E.M. Vogel, L. Colombo, R.S. Ruoff, J. Am. Chem. Soc. **133**, 2816–2819 (2011)
12. D.A.C. Brownson, C.E. Banks, Phys. Chem. Chem. Phys. **14**, 8264–8281 (2012)
13. D.A.C. Brownson, D.K. Kampouris, C.E. Banks, Chem. Soc. Rev. **41**, 6944–6976 (2012)
14. K.K. Cline, M.T. McDermott, R.L. McCreery, J. Phys. Chem. **98**, 5314–5319 (1994)
15. M.C. Henstridge, E. Laborda, N.V. Rees, R.G. Compton, Electrochimica. Acta **84**, 12–20 (2012)
16. R. Nissim, C. Batchelor-McAuley, M.C. Henstridge, R.G. Compton, Chem. Commun. **48**, 3294–3296 (2012)
17. R.L. McCreery, M.T. McDermott, Anal. Chem. **84**, 2602–2605 (2012)
18. N.M.R. Peres, L. Yang, S.-W. Tsai, New J. Phys. **11**, 095007 (2009)
19. D. Jiang, B.G. Sumpter, S. Dai, J. Chem. Phys. **126**, 134701 (2007)
20. M. Pumera, Chem. Soc. Rev. **39**, 4146–4157 (2010)
21. Y. Shimomura, Y. Takane, K. Wakabayashi, J. Phys. Soc. Jpn. **80**, 054710 (2011)
22. R. Sharma, J.H. Baik, C.J. Perera, M.S. Strano, Nano Lett. **10**, 398–405 (2010)
23. C.E. Banks, T.J. Davies, G.G. Wildgoose, R.G. Compton, Chem. Commun., 2005, 829–841
24. K.R. Ward, N.S. Lawrence, R.S. Hartshorne, R.G. Compton, Phys. Chem. Chem. Phys. **14**, 7264–7275 (2012)
25. M.A. Edwards, P. Bertoncello, P.R. Unwin, J. Phys. Chem. C **113**, 9218–9223 (2009)
26. C.G. Williams, M.A. Edwards, A.L. Colley, J.V. Macpherson, P.R. Unwin, Anal. Chem. **81**, 2486–2495 (2009)
27. S.C.S. Lai, A.N. Patel, K. McKelvey, P.R. Unwin, Angew. Chem. Int. Ed. **51**, 5405–5408 (2012)
28. C. Batchelor-McAuley, E. Laborda, M.C. Henstridge, R. Nissim, R.G. Compton, Electrochim. Acta **88**, 895–898 (2013)

29. W. Li, C. Tan, M.A. Lowe, H.D. Abruna, D.C. Ralph, ACS Nano **5**, 2264–2270 (2011)
30. A.T. Valota, I.A. Kinloch, K.S. Novoselov, C. Casiraghi, A. Eckmann, E.W. Hill, R.A.W. Dryfe, ACS Nano **5**, 8809–8815 (2011)
31. F. Banhart, J. Kotakoski, A.V. Krasheninnikov, ACS Nano **5**, 26–41 (2011)
32. R.S. Robinson, K. Sternitzke, M.T. McDermott, R.L. McCreery, J. Electrochem. Soc. **138**, 2412–2418 (1991)
33. D.A.C. Brownson, L.J. Munro, D.K. Kampouris, C.E. Banks, RSC Adv. **1**, 978–988 (2011)
34. J.C. Meyer, C. Kisielowski, R. Erni, M.D. Rossell, M.F. Crommie, A. Zettl, Nano Lett. **8**, 3582–3586 (2008)
35. M.M. Ugeda, I. Brihuega, F. Guinea, J.M. Gomez-Rodriguez, Phys. Rev. Lett. **104**, 096804 (2010)
36. D.A.C. Brownson, L.C.S. Figueiredo-Filho, X. Ji, M. Gomez-Mingot, J. Iniesta, O. Fatibello-Filho, D.K. Kampouris, C.E. Banks, J. Mater. Chem. A **1**, 5962–5972 (2013)
37. D.K. Kampouris, C.E. Banks, Chem. Commun. **46**, 8986–8988 (2010)
38. P.M. Hallam, C.E. Banks, Electrochem. Commun. **13**, 8–11 (2011)
39. A.G. Guell, N. Ebejer, M.E. Snowden, J.V. Macpherson, P.R. Unwin, J. Am. Chem. Soc. **134**, 7258–7261 (2012)
40. C.X. Lim, H.Y. Hoh, P.K. Ang, K.P. Loh, Anal. Chem. **82**, 7387–7393 (2010)
41. X. Ji, C.E. Banks, A. Crossley, R.G. Compton, Chem. Phys. Chem. **7**, 1337–1344 (2006)
42. A. Chou, T. Böcking, N.K. Singh, J.J. Gooding, Chem. Commun., 2005, 842–844
43. L. Tang, Y. Wang, Y. Li, H. Feng, J. Lu, J. Li, Adv. Funct. Mater. **19**, 2782–2789 (2009)
44. S.P. Kumar, R. Manjunatha, C. Nethravathi, G.S. Suresh, M. Rajamathi, T.V. Venkatesha, Electroanalysis **23**, 842–849 (2011)
45. J. Premkumar, S.B. Khoo, J. Electroanal. Chem. **576**, 105–112 (2005)
46. P. Chen, R.L. McCreery, Anal. Chem. **68**, 3958–3965 (1996)
47. P. Chen, M.A. Fryling, R.L. McCreery, Anal. Chem. **67**, 3115–3122 (1995)
48. G.P. Keeley, A. O'Neill, N. McEvoy, N. Peltekis, J.N. Coleman, G.S. Duesberg, J. Mater. Chem. **20**, 7864–7869 (2010)
49. G.P. Keeley, A. O'Neill, M. Holzinger, S. Cosnier, J.N. Coleman, G.S. Duesberg, Phys. Chem. Chem. Phys. **13**, 7747–7750 (2011)
50. A. Ambrosi, A. Bonanni, M. Pumera, Nanoscale **3**, 2256–2260 (2011)
51. H.L. Poh, F. Sanek, A. Ambrosi, G. Zhao, Z. Sofer, M. Pumera, Nanoscale **4**, 3515–3522 (2012)
52. B. Guo, L. Fang, B. Zhang, J.R. Gong, Insciences J. **1**, 80–89 (2011)
53. X. Huang, X. Qi, F. Boey, H. Zhang, Chem. Soc. Rev. **41**, 666–686 (2012)
54. D.A.C. Brownson, J.P. Metters, D.K. Kampouris, C.E. Banks, Electroanalysis **23**, 894–899 (2011)
55. D.A.C. Brownson, C.E. Banks, Electrochem. Commun. **13**, 111–113 (2011)
56. D.A.C. Brownson, C.E. Banks, Analyst **136**, 2084–2089 (2011)
57. D.A.C. Brownson, C.E. Banks, Chem. Commun. **48**, 1425–1427 (2012)
58. C.H.A. Wong, M. Pumera, Electrochem. Commun. **22**, 105–108 (2012)
59. B. Sljukic, C.E. Banks, R.G. Compton, Nano Lett. **6**, 1556–1558 (2006)
60. C.E. Banks, A. Crossley, C. Salter, S.J. Wilkins, R.G. Compton, Angew. Chem. Int. Ed. **45**, 2533–2537 (2006)
61. A. Ambrosi, S.Y. Chee, B. Khezri, R.D. Webster, Z. Sofer, M. Pumera, Angew. Chem. Int. Ed. **51**, 500–503 (2012)
62. A. Ambrosi, C.K. Chua, B. Khezri, Z. Sofer, R.D. Webster, M. Pumera, Proc. Natl. Acad. Sci. USA **109**, 12899–12904 (2012)
63. M. Giovanni, H.L. Poh, A. Ambrosi, G. Zhao, Z. Sofer, F. Sanek, B. Khezri, R.D. Webster, M. Pumera, Nanoscale **4**, 5002–5008 (2012)
64. X. Li, X. Yang, L. Jia, X. Ma, L. Zhu, Electrochem. Commun. **23**, 94–97 (2012)
65. C. Tan, J. Rodriguez-Lopez, J.J. Parks, N.L. Ritzert, D.C. Ralph, H.D. Abruna, ACS Nano **6**, 3070–3079 (2012)
66. H. Wang, T. Maiyalagan, X. Wang, ACS Catal. **2**, 781–794 (2012)

67. L. Gan, D. Zhang, X. Guo, Small **8**, 1326–1330 (2012)
68. Y. Shao, S. Zhang, M.H. Engelhard, G. Li, G. Shao, Y. Wang, J. Liu, I.A. Aksay, Y. Lin, J. Mater. Chem. **20**, 7491–7496 (2010)
69. H. Liu, Y. Liu, D. Zhu, J. Mater. Chem. **21**, 3335–3345 (2011)
70. D. Wei, Y. Liu, Y. Wang, H. Zhang, L. Huang, G. Yu, Nano Lett. **9**, 1752–1758 (2009)
71. S. Mukherjee, T.P. Kaloni, J. Nanopart. Res. **2012**, 14 (1059)
72. L. Zhao, Science **333**, 999–1003 (2011)
73. A. Lherbier, X. Blase, Y.-M. Niquet, F. Triozon, S. Roche, Phys. Rev. Lett. **101**, 036808 (2008)
74. S.U. Lee, R.V. Belosludov, H. Mizuseki, Y. Kawazoe, Small **5**, 1769–1775 (2009)
75. C. Wang, Y. Zhou, L. He, T.-W. Ng, G. Hong, Q.-H. Wu, F. Gao, C.-S. Lee, W. Zhang, Nanoscale **5**, 600–605 (2013)
76. Z. Luo, S. Lim, Z. Tian, J. Shang, L. Lai, B. MacDonald, C. Fu, Z. Shen, T. Yu, J. Lin, J. Mater. Chem. **21**, 8038–8044 (2011)
77. L. Qu, Y. Liu, J.-B. Baek, L. Dai, ACS Nano **4**, 1321–1326 (2010)
78. A.L.M. Reddy, A. Srivastava, S.R. Gowda, H. Gullapalli, M. Dubey, P.M. Ajayan, ACS Nano **4**, 6337–6342 (2010)
79. Z. Jin, J. Yao, C. Kittrell, J.M. Tour, ACS Nano **5**, 4112–4117 (2011)
80. C. Zhang, L. Fu, N. Liu, M. Liu, Y. Wang, Z. Liu, Adv. Mater. **23**, 1020–1024 (2011)
81. D. Deng, X. Pan, L. Yu, Y. Cui, Y. Jiang, J. Qi, W.-X. Li, Q. Fu, X. Ma, Q. Xue, G. Sun, X. Bao, Chem. Mater. **23**, 1188–1193 (2011)
82. L.S. Panchakarla, K.S. Subrahmanyam, S.K. Saha, A. Govindaraj, H.R. Krishnamurthy, U.V. Waghmare, C.N.R. Rao, Adv. Mater. **21**, 4726–4730 (2009)
83. A. Ghosh, D.J. Late, L.S. Panchakarla, A. Govindaraj, C.N.R. Rao, J. Exp. Nanosci. **4**, 313–322 (2009)
84. B. Guo, Q. Liu, E. Chen, H. Zhu, L. Fang, J.R. Gong, Nano Lett. **10**, 4975–4980 (2010)
85. D. Geng, Y. Chen, Y. Chen, Y. Li, R. Li, X. Sun, S. Ye, S. Knights, Energy Environ. Sci. **4**, 760–764 (2011)
86. X. Wang, X. Li, L. Zhang, Y. Yoon, P.K. Weber, H. Wang, J. Guo, H. Dai, Science **324**, 768–771 (2009)
87. X. Li, H. Wang, J.T. Robinson, H. Sanchez, G. Diankov, H.J. Dai, J. Am. Chem. Soc. **131**, 15939–15944 (2009)
88. L.S. Zhang, X.Q. Liang, W.G. Song, Z.Y. Wu, Phys. Chem. Chem. Phys. **12**, 12055–12059 (2010)
89. Z.H. Sheng, L. Shao, J.J. Chen, W.J. Bao, F.B. Wang, X.H. Xia, ACS Nano **5**, 4350–4358 (2011)
90. R.I. Jafri, N. Rajalakshmi, S. Ramaprabhu, J. Mater. Chem. **20**, 7114–7117 (2010)
91. Y. Wang, Y. Shao, D.W. Matson, J. Li, Y. Lin, ACS Nano **4**, 1790–1798 (2010)
92. H.M. Jeong, J.W. Lee, W.H. Shin, Y.J. Choi, H.J. Shin, J.K. Kang, J.W. Choi, Nano Lett. **11**, 2472–2477 (2011)
93. Y.C. Lin, C.Y. Lin, P.W. Chiu, Appl. Phys. Lett. **96**, 133110 (2010)
94. D. Long, W. Li, L. Ling, J. Miyawaki, I. Mochida, S.H. Yoon, Langmuir **26**, 16096–16102 (2010)
95. D.W. Wang, I.R. Gentle, G.Q. Lu, Electrochem. Commun. **12**, 1423–1427 (2010)
96. C.P. Ewels, M. Glerup, J. Nanosci. Nanotechnol. **5**, 1345–1363 (2005)
97. J. Casanovas, J.M. Ricart, J. Rubio, F. Illas, J.M. Jiménez-Mateos, J. Am. Chem. Soc. **118**, 8071–8076 (1996)
98. Z. Lin, M.-K. Song, Y. Ding, Y. Liu, M. Liu, C.-P. Wong, Phys. Chem. Chem. Phys. **14**, 3381–3387 (2012)
99. G.-X. Ma, J.-H. Zhao, J.-F. Zheng, Z.-P. Zhu, Carbon **51**, 435 (2013)
100. K. Parvez, S. Yang, Y. Hernandez, A. Winter, A. Turchanin, X. Feng, K. Mullen, ACS Nano **6**, 9541–9550 (2012)
101. B. Zheng, J. Wang, F.-B. Wang, X.-H. Xia, Electrochem. Commun. **28**, 24–26 (2013)
102. D.R. Dreyer, S. Park, C.W. Bielawski, R.S. Rouff, Chem. Soc. Rev. **39**, 228–240 (2010)

103. L.J. Cote, J. Kim, V.C. Tung, J. Luo, F. Kim, J. Huang, Pure Appl. Chem. **83**, 95–110 (2011)
104. T. Szabo, O. Berkesi, P. Forgo, K. Josepovits, Y. Sanakis, D. Petridis, I. Dekany, Chem. Mater. **18**, 2740–2749 (2006)
105. M. Zhou, Y. Wang, Y. Zhai, J. Zhai, W. Ren, F. Wang, S. Dong, Chem. Eur. J. **15**, 6116–6120 (2009)
106. D.A.C. Brownson, A.C. Lacombe, M. Gomez-Mingot, C.E. Banks, RSC Adv. **2**, 665–668 (2012)
107. D.A.C. Brownson, C.E. Banks, Phys. Chem. Chem. Phys. **13**, 15825–15828 (2011)
108. D.A.C. Brownson, M. Gomez-Mingot, C.E. Banks, Phys. Chem. Chem. Phys. **13**, 20284–20288 (2011)
109. A.N. Obraztsov, E.A. Obraztsova, A.V. Tyurnina, A.A. Zolotukhin, Carbon **45**, 2017–2021 (2007)
110. K.S. Kim, Y. Zhao, H. Jang, S.Y. Lee, J.M. Kim, K.S. Kim, J.-H. Ahn, P. Kim, J.-Y. Choi, B.H. Hong, Nature **457**, 706–710 (2009)
111. A. Guermoune, T. Chari, F. Popescu, S.S. Sabri, J. Guillemette, H.S. Skulason, T. Szkopek, M. Siaj, Carbon **49**, 4204–4210 (2011)
112. A. Ambrosi, A. Bonanni, Z. Sofer, M. Pumera, Nanoscale **5**, 2379–2387 (2013)

Chapter 4
Graphene Applications

> *Not everything that can be counted counts, and not everything that counts can be counted.*
>
> William Bruce Cameron

In the earlier chapters we have observed the various electrochemical studies that aim to extend our fundamental knowledge of graphene electrochemistry. An area which greatly supports this is the use of graphene as an electrode material in the field of electroanalysis and also in energy storage/conversion. Considering electroanalysis for example, it is readily evident that electrochemistry has widespread implications in the fabrication of sensing devices that have global ramifications in terms of the detection of substances harmful to human health and the environment. One of the most successful commercialisation routes whilst using electrochemistry is the 'billion dollar per annum' glucose sensing market, allowing diabetics to instantly measure their blood glucose without having to travel to a clinic or hospital [1]. Typically, such sensors are based upon carbon/graphite derivatives and are fabricated via screen-printing methods due to the advantages possessed in terms of mass production, where economical yet reproducible sensors can be produced [2]. Other sensors based on this format are being commercially developed.

In terms of the performance obtained from a given electrochemical device, it is the properties of the electrode material itself that are most significant. Therefore research is conducted into the utilisation of various electrode materials, such as graphene. As mentioned above, graphitic forms of carbon in-particular have widespread use as disposable electrode materials due to their relatively economical and scale-up fabrication processes [2], in addition to their non-toxic and highly conductive attributes. Graphene is no exception to this trend, where owing to the combination of its interesting electrochemical properties (Chap. 3) and it's reported unique assortment of physiochemical properties (Chap. 1) it has enormous potential to be beneficially utilised in a range of sensing and energy related electrochemical applications.

Indeed, probably one of the very reasons you are reading this book is that you have read that graphene is a fantastic material for use in electroanalysis and in energy related electrochemical applications. On inspection of the literature one would believe that this is true and we could endlessly list the reported papers where graphene modified electrodes have been reported to be useful in electroanalysis, or have provided benefits in energy storage and generation applications. However, upon closer inspection of this literature, all is not what it seems and we aim to

D. A. C. Brownson and C. E. Banks, *The Handbook of Graphene Electrochemistry*, DOI: 10.1007/978-1-4471-6428-9_4, © Springer-Verlag London Ltd. 2014

provide a concise and informative overview of the field of graphene use in electroanalysis and additionally where graphene has been explored in energy storage
(supercapacitors and so on) and conversion (fuel cells etc.); offering constructive
insights into this exciting field and allowing readers to make their own mind up on
whether graphene has revolutionised or is yet to revolutionise these fields.

4.1 Sensing Applications of Graphene

4.1.1 Electroanalysis

A common issue faced by graphene experimentalists is 'how to electrically wire/
connect to graphene' such that one can obtain the reported benefits. A common
approach in the field which solves this problem is to modify a pre-existing/well
utilised electrode material, such as glassy carbon (GC), boron-doped diamond
(BDD), and screen-printed electrodes; and in lesser cases, edge plane and basal
plane pyrolytic graphite (EPPG and BPPG respectively) electrodes. Typically the
chosen electrode substrate is modified with aliquots (usually μL volumes) of a
graphene suspension/solution and as such, one can control the surface coverage
and study this as a function of the electroanalytical response.

Two of the very first examples of graphene being utilised in electroanalysis were
towards the sensing of dopamine [3] and β-nicotinamide adenine dinucleotide
(NADH) [4]. In the former case, graphene modified GC electrodes were shown to
allow the sensing of dopamine in the presence of ascorbic acid, which typically
coexists with dopamine in biological organisms and shares a similar oxidation
potential for electrochemical detection. The authors reported that the graphene-
modified electrode completely eliminated the issue of the coexisting (intervening)
oxidation peaks. Furthermore, when contrasted with MWCNTs, graphene was
found to yield an improved electrochemical response, which the authors suggested
was due to graphene's unique electronic structure. For the case of the distinct
voltammetric responses at dopamine over ascorbic acid (which were negligible and
eliminated) the authors attributed this to $\pi-\pi$ interactions between the phenyl
structure of dopamine and the two-dimensional planar hexagonal carbon structure of
graphene, suggesting that this makes electron transfer feasible, while ascorbic acid
oxidation is inactive, suggested to be most likely because of its weak $\pi-\pi$ interaction
with graphene [3]. It was demonstrated that dopamine could be measured over the
range of 5 to 200 μM in a large excess of ascorbic acid, which is extremely useful
for potential implementation of graphene into real sensing applications.

In terms of the latter case above [4], cyclic voltammograms obtained using
graphene modified BPPG and EPPG electrodes in a 0.1 M phosphate buffer
solution (PBS, pH 7) containing 2 mM NADH are shown in Fig. 4.1. The authors
reported that for the electrochemical oxidation of NADH the graphene modified
electrodes exhibited a substantial negative shift of the anodic peak potential
(an improved electrochemical response) when compared with bare/unmodified

Fig. 4.1 Cyclic voltammograms obtained using (*a*) bare BPPG electrode (*b*) graphene modified BPPG electrode (*c*) bare EPPG electrode, and (*d*) graphene modified EPPG electrode in 0.1 M phosphate buffer solution (PBS, pH 7) containing 2 mM NADH. Reproduced from Ref. [4] with permission from Elsevier

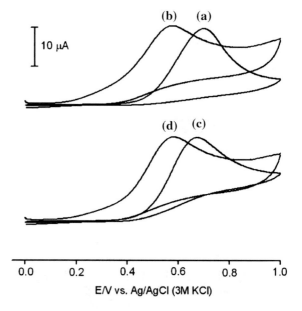

BPPG and EPPG electrodes. The authors suggest that the "electro-catalytic" (our emphasis) oxidation of NADH when utilising graphene is evident from the oxidation peak potentials occurring at ~ 0.564 and 0.652 V when using a graphene modified EPPG and an unmodified EPPG electrode respectively, with the authors suggesting that a decrease of activation energy at the graphene modified BPPG and EPPG electrodes allows the detection of NADH at lower potentials, which might be attributed to the unique electronic structure and properties of graphene [4].

Unfortunately the authors [4] have been caught up in the graphene gold rush and it is clear from inspection of Fig. 4.1 that the bare BPPG and EPPG electrodes utilised in their work give rise to the same oxidation potentials for NADH—when in reality the EPPG response is significantly more reversible that that presented in Fig. 4.1; interested readers should consult Ref. [5].

In other work, the electrochemical detection of paracetamol has been reported [6]; the evidence is presented in Fig. 4.2. The authors reported that at the bare GC electrode, that is, prior to being decorated with graphene (graphene sheets functionalised with hydroxyl and carboxylic groups), the electrochemical oxidation of paracetamol exhibits electrochemically irreversible behaviour with relatively weak redox peaks and currents at an E_{pa} (anodic peak potential) of ~ 0.368 V and a E_{pc} (cathodic peak potential) of ~ 0.101 V. However, with the addition of graphene, the modified electrode exhibits a pair of well-defined redox peaks towards paracetamol, with an E_{pa} of ~ 0.273 V and a E_{pc} of ~ 0.231 V; a reduction in the overpotential of paracetamol is evident when contrasting the graphene modified electrode to that of the unmodified GC electrode, with a shift of 95 mV [6]. The unique change (see Fig. 4.2) is attributed to the nano-composite film of graphene accelerating the electrochemical reaction [6]. The effect of scan rates on the redox

Fig. 4.2 Cyclic
voltammograms recorded at a
bare GC electrode (*a*) with
100 μM paracetamol;
graphene/GC electrode with
(*b*) 20 μM paracetamol and
without paracetamol (*c*) in the
buffer of 0.1 M NH$_3$·H$_2$O–
NH$_4$Cl, pH 9.3, scan rate:
50 mV s^{-1}. Reproduced from
Ref. [6] with permission from
Elsevier

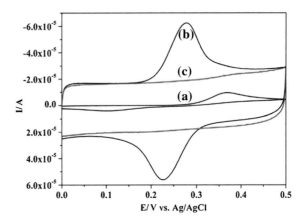

couple of paracetamol at the graphene-modified GC electrode was investigated by
cyclic voltammetry, which indicated that the modified-electrode reaction of paracetamol was a surface-confined process [6]. Note however, this could be due to a
change in mass transport arising from the graphene film being potentially porous,
where thin-layer type behaviour is observed (see Sect. 2.11); this behaviour corresponds to the graphene electrochemical response being in *Zone III* (see Chap. 3).
In this case it is likely that the large background current exhibited by the graphene
modified electrode is due to the large accessible surface area of such a porous
surface. Nevertheless, the authors went on to show that graphene modified electrode allowed a detection limit of 3.2×10^{-8} M, a reproducibility of 5.2 %
(relative standard deviation), and a satisfied recovery ranging from 96.4 to 103.
3 %. Note that Appendix B provides a useful overview of how to analyse data and
perform such useful analytical benchmarks that are widely used in the literature.

Other curious work has reported using reduced graphene sheets for the sensing
of kojic acid, which is a natural metabolic product of several species of *Aspergillus*,
Acetobacter, and *Penicillium* [7]. Additionally, kojic acid and some of its derivatives are used in cosmetic preparation, to achieve a skin-lightening effect and also
as a food additive and preservative. Due to potential carcinogenic properties, the
development of a convenient, economical, rapid and sensitive method for the
determination of trace amounts of kojic acid in different samples is reported to be
highly desirable [7]. The cyclic voltammetric responses at bare GC and graphite
electrodes (in 0.2 M HAc–NaAc solution, pH 6.0) in the presence and absence of
200 μM kojic acid are shown in Fig. 4.3. Note that the graphite electrode used in
this work [7] is not defined as to whether this is an EPPG or BPPG electrode of
HOPG, or of other form. The response at the bare GC and graphite electrodes
(curves b and d, respectively) indicate that while kojic acid can be electro-oxidised
the voltammetric peaks are not well-defined. However the authors show that if the
GC electrode is modified with graphene (synthesised via the reduction of GO, i.e.
reduced graphene sheets (RGSs)) a well-identified anodic peak is observed, with
the peak potential (E_p) decreasing to ~ 0.87 V (curve f) and a significant increment

Fig. 4.3 Cyclic voltammograms at (*a* and *b*) the unmodified/bare GC electrode (*c* and *d*) the graphite electrode, and (*e* and *f*) the reduced-graphene-sheet/GC electrode in 0.2 M HAc–NaAc solution (pH 6.0) (*a*, *c*, and *e*) in the absence and (*b*, *d*, and *f*) in the presence of 200 μM kojic acid. Scan rate: 100 mV s^{-1}. Reproduced from Ref. [7] with permission from Elsevier

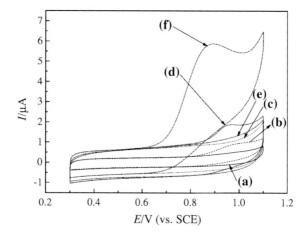

in the peak current evident; leading the authors to suggest that the RGSs have a high electro-catalytic activity towards the oxidation of kojic acid [7]. The authors go on further in their paper to suggest that the decrease in over-potential and large enhancement of the oxidation current for kojic acid when using the RGSs/GC electrode can be attributed to the high edge-plane sites/defects of RGSs.

It is evident that the "electro-catalytic" behaviour and enhanced analytical performance of graphene and related structures has been widely reported, encompassing the detection of a diverse range of analytes including numerous bio-molecules, gases and miscellaneous organic and inorganic compounds [8, 9], and we expect this trend to extend. However, it is interesting to note that in the above examples (and in many others) the electro-catalysis of graphene is compared only to the underlying electrode (usually GC) and not routinely to other relevant graphitic materials, such as HOPG and indeed that of graphite. This is a significant issue given that in the majority of studies the graphene utilised is most probably *quasi*-graphene [10], that Is, consisting of multi-layers and theoretically approaching the structural composition of *graphite*. In the above specific example presented in Fig. 4.3, the authors do indeed try and compare their response to that of an (undefined) graphitic electrode and show that their modification with RGSs gives rise to useful voltammetry [7]. However, the authors failed to perform appropriate control experiments with the *starting material*, that is, graphene oxide (GO) and additionally, simply graphite modified GC electrodes; the role of oxygenated species that may reside on the reduced GO is not clear and likely may contribute significantly to the observed "electro-catalytic" response reported in Fig. 4.3.

Additionally, the beneficial response of using graphene modified GC electrodes towards the sensing of paracetamol (as highlighted above) needs to be considered further. The authors suggest that improvements are due to the "graphene accelerating the electrochemical reaction due to the defective sheets of graphene" [6]. The lack of control experiments means that the true origin of the claimed 'electro-catalysis' is unclear given that the graphene sheets utilised in their work are functionalised with hydroxyl and carboxylic groups, these may contribute to the response observed (see Fig. 4.2). Alternatively, the response may arise through graphitic impurities formed as a result of the fabrication process, which also can contribute to a highly disordered and porous graphene structure—in such a case it is likely that the observed response is due to 'thin-layer' effects (see Sect. 2.11).

Note that the majority of graphene used in electrochemistry is produced through the reduction of GO, which results in partially functionalised graphene sheets or chemically reduced GO, which has abundant structural defects (edge plane like-sites/defects) [9, 11] with carbon-oxygen functional groups that are remaining after this fabrication process and have been shown to greatly influence electrochemical activity; which can be either advantageous or detrimental towards the sensing of a target analyte [8, 12–15].

While, as we write this book, there are many encouraging examples of the reported benefits of utilising graphene in electroanalytical applications, where improved sensitivities and low detection limits have been reported, it is still evident that in the majority of cases the electrochemical performance of *true* graphene is scarcely reported and it is vital that appropriate control and comparison experiments are performed prior to the beneficial electro-catalysis of graphene being reported; Appendix C provides a useful (non-exhaustive) summary of the various control experiments that should be considered.

In the case of *true* graphene modified electrodes, that is, the utilisation of pristine graphene, the reported (above) beneficial electroanalytical responses have been explored towards the electroanalytical sensing of dopamine (DA), uric acid (UA), paracetamol (AP) and p-benzoquinone (BQ) [16]. Note that to allow the true electroanalytical capabilities of graphene to be de-convoluted the authors chose to modify HOPG electrode substrates (so a material of comparable nature to that of graphene) that exhibited either fast or slow (EPPG or BPPG respectively) electron transfer properties (i.e. either favourable or undesirable electrochemical properties, respectively). Figure 4.4 depicts the comparison of the response observed at different electrode materials and at pristine graphene modified electrodes towards the electroanalytical sensing of DA, where a large cyclic voltammetric signature is observed using the bare EPPG electrode, which is consistent with the literature. Also shown is the response of the bare BPPG electrode where a cyclic voltammetric response is observed with a larger peak-to-peak (ΔE_P) separation; such a response is consistent with the difference in surface structure where the EPPG has a greater coverage of useful edge plane- like sites/defects [16].

Interestingly, the introduction of pristine graphene upon both of the electrode substrates resulted in a decrease in the voltammetric response (in both cases), with reductions in the peak heights (the analytically important current) and increases in the ΔE_P separation (indicating slow, unfavourable, heterogeneous electron transfer kinetics) [16]. Note that the coverage of graphene utilised was chosen specifically as this was found to correspond to a near true graphene modified electrode. You will recall from Chap. 3 that three 'zones' are evident in graphene electrochemistry, where in *zone I* the coverage corresponds to the use of graphene modified electrodes where the graphene is near to that, as can be achieved, of mono- and few-layer

Fig. 4.4 Cyclic voltammetric profiles recorded for 50 μM dopamine (DA) in pH 7 PBS using unmodified EPPG (*solid line*) and BPPG (*dot-dashed line*) electrodes, and 20 ng pristine graphene modified EPPG (*dashed line*) and BPPG (*dotted line*) electrodes. Scan rate: 100 mVs^{-1} (vs. SCE). Reproduced from Ref. [16] with permission from The Royal Society of Chemistry

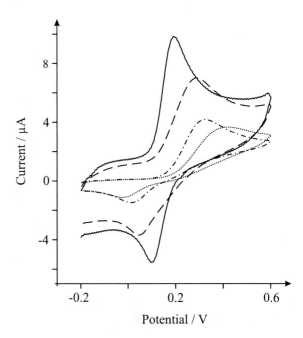

graphene; which is as that presented in Fig. 4.4. A good question to ask is, *why not keep adding more graphene until an improvement is observed?* Again, this change in coverage will result in the modified electrode being within *zone II*, where multi-layer graphene will occur and hence why use graphene at all, since you are recreating the structural composition of graphite! Such a naive question shows exactly why control experiments are conveniently avoided within the literature.

Returning to the case of exploring the electroanalytical response of pristine graphene modified electrodes, Fig. 4.5 depicts the response of making additions of dopamine into a buffer solution, using graphene modified electrodes along with a calibration plot arising from using the different electrode materials. Strikingly the best response, that is the response with the greatest gradient and hence sensitivity, corresponds to the case of a bare EPPG electrode. Indeed Table 4.1 summaries the electroanalytical responses observed towards different analytical targets using the different electrode materials, with their corresponding sensitivities and limit of detections (LOD) noted [16].

It is clearly evident that through the introduction of graphene onto the target electrode substrates, there is a reduced electroanalytical performance at the pristine graphene compared to the bare/unmodified electrodes; which is contradictory to the majority of current literature reports which claim that the application of graphene leads to an enhancement in the electroanalytical response in many instances. [6, 7, 17–21] However importantly, such poor performances were evident at both the EPPG (fast electrode kinetics) and BPPG (slow electrode kinetics) graphene modified electrodes and, in accordance with fundamental theory (see Chap. 3), is as expected given the insights gained from earlier work utilising

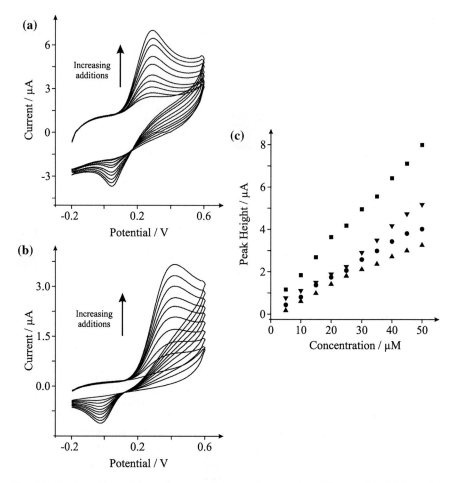

Fig. 4.5 Cyclic voltammetric profiles recorded towards successive additions of 5 μM dopamine (DA) into a pH 7 PBS (5–50 μM) utilising EPPG (**a**) and BPPG (**b**) electrodes following modification with 20 ng graphene. **c** Calibration plots towards the detection of DA depicting the peak height as a function of concentration, obtained via cyclic voltammetric measurements performed using unmodified EPPG (*squares*) and BPPG (*circles*) electrodes in addition to EPPG (*inverted triangles*) and BPPG (*triangles*) electrodes following modification with 20 ng graphene. All data obtained at a scan rate of 100 mVs^{-1} (vs. SCE). Reproduced from Ref. [16] with permission from The Royal Society of Chemistry

identical pristine graphene modified graphitic electrodes [22]. As mentioned in Chap. 3 pristine graphene has a high electron density around its edge as opposed to its centre, where owing to its unique geometry graphene possesses a low portion of edge plane like- surface area and resultantly exhibits slow heterogeneous electron transfer kinetics, particularly when compared to its closest counterpart—graphite [22]. Thus the response observed above can be attributed (in both cases) to the relative coverage of available edge plane sites (electroactive sites) on a given

Table 4.1 Comparison of the analytical sensitivities and resultant LODs (based on three-sigma) obtained at the various electrode materials/modifications towards the electroanalysis of dopamine (DA), uric acid (UA), paracetamol (AP) and *p*-benzoquinone (BQ) (20 ng graphene modification for DA, UA and AP, 40 ng modification for BQ) ($N = 3$)

Electrode material	Sensitivity/A M^{-1}	LOD (3σ)/μM
DA		
EPPG	0.15	1.73 ($\pm$0.03)
Graphene/EPPG	0.10	3.78 ($\pm$0.08)
BPPG	0.08	2.44 ($\pm$0.05)
Graphene/BPPG	0.07	4.18 ($\pm$0.11)
UA		
EPPG	0.13	10.40 ($\pm$0.48)
Graphene/EPPG	0.09	11.25 ($\pm$0.40)
BPPG	0.08	11.18 ($\pm$0.36)
Graphene/BPPG	0.04	14.02 ($\pm$0.46)
AP		
EPPG	0.21	2.41 ($\pm$0.06)
Graphene/EPPG	0.14	2.77 ($\pm$0.08)
BPPG	0.12	3.33 ($\pm$0.11)
Graphene/BPPG	0.10	4.11 ($\pm$0.14)
BQ		
EPPG	−0.14	2.14 ($\pm$0.05)
Graphene/EPPG	−0.10	3.05 ($\pm$0.04)
BPPG	−0.08	3.40 ($\pm$0.15)
Graphene/BPPG	−0.08	2.31 ($\pm$0.09)

Reproduced from Ref. [16] with permission from The Royal Society of Chemistry

electrode surface, where through the introduction of graphene the number of these sites is significantly reduced, which are replaced instead with the relatively inactive pristine basal plane sites of graphene which result in a 'blocked' electrode surface, and hence reduced electron transfer kinetics give rise to poor electrochemical characteristics. As a result a change in the overall reversibility/irreversibility of the electrode kinetics is observed [16, 23, 24]; note that where large deviations in the analytical signal occur it is inferred that this is due to the inherent lack of oxygenated species on the pristine graphene that are readily present upon HOPG surfaces (depending on pre-treatment of the electrode surface) and are beneficial in some instances [16, 25, 26].

Other work has considered pristine graphene modified (graphite based) screen-printed electrodes (SPEs) towards the electroanalytical detection of cadmium(II) via anodic stripping voltammetry [27]. Figure 4.6a shows the effect of varying the mass (coverage) of graphene immobilised onto the SPE surface; where, as the coverage of graphene is increased, the electroanalytical parameter that is the voltammetric stripping peak, exhibits a reduction in magnitude. Shown in Fig. 4.6b is the electroanalytical response of the graphene modified SPEs towards the sensing of cadmium(II) where additionally a control experiment using a bare (graphene free) SPE is also shown [27].

It is readily evident that there is no real improvement in the electrochemical response from utilising graphene, and indeed the introduction of graphene upon

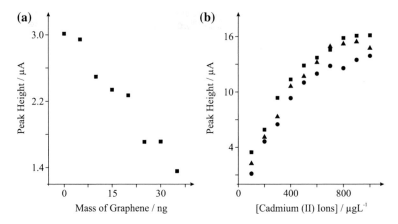

Fig. 4.6 a Relationship between the mass of graphene coverage on the electrode surface and the resultant peak height towards the detection of 400 μgL^{-1} cadmium(II) ions in pH 1.5 HCl aqueous solution. **b** Calibration plots indicating the relationship between the concentration of cadmium(II) ions and the observed peak height at an unmodified SPE (*squares*) and at SPEs following modification with 20 (*triangles*) and 35 ng (*circles*) of graphene. **a** and **b** were both obtained via square-wave anodic stripping voltammetry utilising a deposition potential of −1.2 V (vs. SCE) for 120 s. Reproduced from Ref. [27] with permission from The Royal Society of Chemistry

the electrode surface resulted, in this case, in a reduction in the observed voltammetric signal—leading a to decreased analytical sensitivity towards the detection of cadmium(II) ions at graphene modified SPEs. Modification with graphene also resulted in the reduced inter-reproducibility of the analytical signal, where the % Relative Standard Deviation (%RSD) of the observed peak height/area at an unmodified SPE and a SPE following modification with 20 and 35 ng of graphene were 0.7/4.4, 5.2/2.8 and 16.3/15.8 % (N = 4) respectively (towards the detection of 400 μgL^{-1} (400 ppb) cadmium(II) ions. Note again the coverages (mass additions) utilised were chosen in order to be akin to that of *zone I* graphene electrochemistry [27].

Thus, as it is well known that metals nucleate exclusively on the edge plane sites of graphitic materials, the low edge to basal surface area ratio of the graphene structure of pristine graphene (with virtually no defects across the graphene basal surface) results in poor voltammetric responses; this coupled with the reproducibility at graphene modified electrodes being above the typically accepted RSD of 5 % (thus generally analytically unacceptable), suggests that the use of pristine graphene in the electroanalysis of metal ions via anodic stripping voltammetry is not beneficial [27].

In additional support of the above two studies [16, 27], this same trend was also found to be the case in earlier work towards the detection of hydrogen peroxide at surfactant adsorbed/contaminated graphene modified electrodes [28]. Through careful control experimentation utilising various surfactant and graphite modified electrodes it was concluded that the graphite modified electrode exhibited a

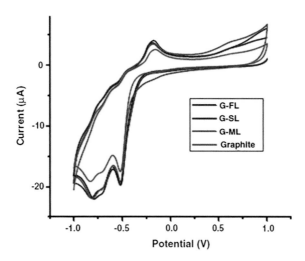

Fig. 4.7 Cyclic voltammograms of 14 μg/mL TNT on single- (*G-SL*), few- (*G-FL*), and multilayer (*G-ML*) graphene nanoribbons and graphite microparticles. Conditions: background electrolyte, 0.5 M NaCl; scan rate, 100 mV s^{-1}. Reproduced from Ref. [21] with permission from Springer Science and Business Media

superior electrochemical response due to its enhanced percentage of edge plane sites when compared to that of graphene; sensitivities were measured at 32.8 and 49.0 μA mM^{-1} for graphene and graphite modified electrodes respectively [28]. Interestingly however, when Nafion^{TM}, routinely used in amperometric biosensors, was introduced onto the graphene and graphite modified electrodes, re-orientation was observed to occur in both cases which proved beneficial in the former and detrimental in the latter (due to alterations in the availability/accessibility of each materials respective edge plane sites) [28].

Other useful work [21, 29] has emerged demonstrating that single-, few-, and multi-layer graphenes modified upon GC electrodes do not appear to give rise to significant advantages over control experiments using graphite micro-particles in terms of the observed sensitivity, linearity and repeatability towards the electro-analytical detection of UA and *L*-ascorbic acid (AA). These observations were extended with regards to the electrochemical detection of 2,4,6-trinitrotoluene (TNT). Figure 4.7 shows the cyclic voltammetric responses obtained from using single-, few-, and multi-layer graphenes modified upon a GC electrode for the electrochemical monitoring of TNT. The reduction of the three nitro-groups of TNT results in three reduction waves at similar potentials of ~ -525, -706 and -810 mV, reflecting stepwise reductions of the individual nitro groups of TNT. Figure 4.8 compares the analytical response of the different graphenes in terms of their calibration plots towards TNT. It is evident that a graphite-microparticle modified electrode provides slight improvements in terms of the electroanalytical sensitivity when compared to the performance of single-, few-, and multi-layer graphene alternatives. In fact, we suggest that it is evident that the single layer

Fig. 4.8 Concentration dependence of TNT on *G-SL*, *G-FL*, *G-ML* (see Fig. 4.7), and graphite electrode surfaces in 0.5 M NaCl using differential pulse voltammetry. Reproduced from Ref. [21] with permission from Springer Science and Business Media

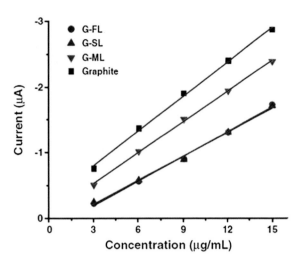

modified electrode gives rise to the lowest response while that of the graphite and multi-layer graphene are quite similar. This is consistent with the concept that single layer graphene exhibits a poor electrochemical response and that as the number of graphene layers are increased, towards that of graphite, an improved response is observed due to the greater proportion of edge plane sites.

In summary, from the current literature and the work highlighted above, it is evident that due to graphene's structural composition and consequently slow heterogeneous electron transfer rates, there appears to be no real advantages of employing *pristine graphene* in an electroanalytical context where fast heterogeneous electron transfer properties are desirable, as covered here. It is noted that within these studies the EPPG electrode consistently exhibited the best performance owing to its favourable orientation of the graphite planes resulting in optimum edge plane coverage (and accessibility), and in certain cases the presence of oxygenated species, [16, 22, 27, 30, 31] which questions the need to modify electrodes with pristine graphene for the electroanalysis of target analytes in the first place.

Given the wide range of graphenes that are available, which can significantly differ from each other since different methods of preparation are used, control experiments need to be routinely performed; Appendix C provides a useful summary for the graphene experimentalist.

As discussed above, typically one modifies an electrode substrate with aliquots of a graphene suspension/solution and as such, one can control the surface coverage and study this as a function of the electroanalytical response. However, since graphene prefers its lowest energy confirmation, that is, bulk graphene, viz graphite, aggregation of graphene sheets occurs and effectively one has a surface consisting of, in the best case, *quasi*-graphene [10] and through additions, deviation occurs to the extent that the surface is essentially consisting of significant layers of graphene, viz graphite. An approach to overcome aggregation is to have a true graphene electrode which can be readily fabricated via Chemical Vapour Deposition (CVD), where the graphene is effectively pre-prepared onto a substrate and characterisation of the surface allows one to know exactly the quality of graphene being utilised.

As such the electroanalytical performance of a commercially available CVD grown graphene electrode has been explored towards the sensing of the biologically important analytes, NADH and UA [31]. Figure 4.9 depicts AFM images of the graphene surface which has been grown via CVD upon a nickel film. The image shows a highly disordered surface with graphene orientated both parallel and perpendicular to the surface in addition to the prevalence of graphitic islands across its surface which contributes to a large global coverage of edge plane sites; note that graphene surfaces are known to be polycrystalline, highly disordered and defect abundant when grown on Nickel substrates utilising CVD [32]. Utilising cyclic voltammetry the analytical sensitivities observed towards the detection of NADH at CVD graphene, EPPG and BPPG electrodes were found to correspond to 0.26, 0.22 and 0.15 $Acm^{-2}M^{-1}$ respectively [31]. The similar responses observed at the EPPG and CVD graphene electrodes was inferred to be due to the similar electronic structures of the electrode surfaces due to similar structures where both exhibit a good proportion of edge plane like- sited/defects; the enhanced performance of the CVD graphene over that of the BPPG electrode is as expected given the former possesses a greater degree of edge plane sites in comparison to the latter (as would an EPPG electrode). In the case of UA the analytical sensitivities of the CVD graphene, EPPG and BPPG electrode were found to correspond to 0.48, 0.61 and 0.33 Acm^{-2} M^{-1} respectively, where notably although the performance of the CVD graphene electrode surpassed that of the BPPG electrode, its performance was inferior to that observed at the EPPG electrode, which was inferred to be due to UA's respective surface sensitivity towards the (favourable) presence of oxygenated species and thus was indicative of the differing O/C ratios at the two electrodes [30, 31].

Other useful work has reported the beneficial implementation of an electrochemically anodised CVD grown graphene electrode towards the electroanalysis of various nucleic acids, uric acid (UA), dopamine (DA) and ascorbic acid (AA). Anodising the electrode surface increases the degree of oxygen-related edge plane defects and results in improved electron transfer kinetics being exhibited over that of pristine graphene. Utilising differential pulse voltammetry (DPV) it was demonstrated that mixtures of nucleic acids (adenine, thymine, cytosine and guanine) or bio-molecules (AA, UA and DA) can be resolved as individual peaks and that the anodised graphene electrode could simultaneously detect all four DNA bases in double stranded DNA (dsDNA) without a pre-hydrolysis step, in addition to differentiating single stranded DNA from dsDNA [33]. Of note is the control experiments performed by the authors, where the response of the anodised graphene electrode was compared to that of pristine graphene, GC and BDD alternatives; the response of the anodised graphene electrode was found to surpass and out-perform each alternative [33]. This work shows that graphene with an elevated level of edge plane defects, as opposed to pristine graphene, is the choice platform in electrochemical sensing.

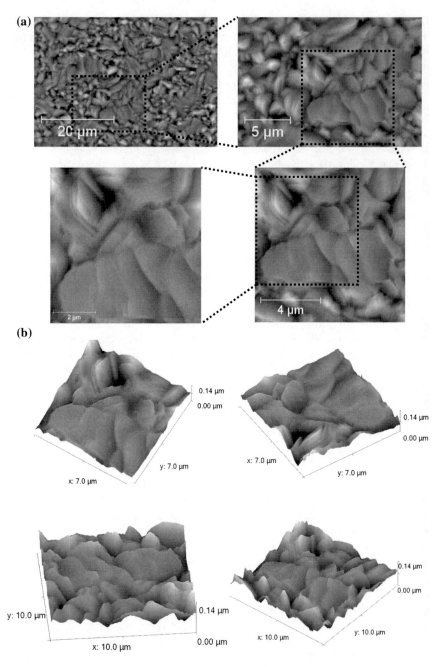

Fig. 4.9 AFM images of the 'as received' commercially available CVD graphene surface, observed from various *top-down* (**a**) and three-dimensional (**b**) perspectives. Note that the surface is generally akin to that of a HOPG surface, but is highly disordered and thus as a result there is a high edge plane content making, the two electrodes in comparison act electrochemically similar. Reproduced from Ref. [31] with permission from The Royal Society of Chemistry

In support of this work there are various reports where graphene, that is highly disordered or defect abundant (high edge plane content) or of which possesses a beneficial oxygen content, has been utilised advantageously within select sensing applications [34–36]. Interestingly, while the functionalisation of graphene is deleterious to its electrical conductivity, the resulting oxygen-containing groups and structural defects can be beneficial for electrochemistry as these are the major sites for rapid heterogeneous electron transfer at some surface sensitive analytes/ probes: note that these sites also provide convenient attachment sites in the development of nanoarchitectonics or the attachment/adsorption of bio-molecules where specific groups can be introduced that play vital roles in electrochemical sensing and energy applications, hence the electrochemical properties of graphene-based electrodes can be modified or tuned by chemical modification and tailored to suit appropriate applications [37].

Thus it is clear that for the analytes studied above, in the best case, the electroanalytical performance of the CVD graphene electrode mimicked that of EPPG, again suggesting no significant advantage of utilising CVD graphene in this analytical context [31]. It is evident that if true graphene is employed in the context covered above (where fast electron transfer is beneficial), the best response that can be achieved is that akin to an EPPG electrode, that is, as governed by the Randles–Ševćik equations; any deviation from this results from either thin-layer behaviour or could possibly be due to changes in mass transport characteristics or other contributing factors (such as impurities, see Sects. 3.2.2 and 3.2.3) rather than (wrongly) assigning the graphene to be 'electro-catalytic'.

Apart from developing a graphene material which has a high density of edge plane like- sites/defects, without recourse to graphite, and with favourable oxygenated species (for the target analyte of interest)—*what useful properties of graphene can be beneficially utilised*?

In the case of pristine graphene, the large proportion of basal plane sites would give rise to many adsorption sites for target analytes. Indeed this has been shown to be the case for pristine graphene towards the detection of the DNA bases, adenine and guanine, where the former adsorbs onto basal plane sites while edge plane sites are electron transfer sites; confirmation was provided through the use of graphitic (EPPG and BPPG) electrodes as control measures [38]. However, due to the low proportion of edge plane sites arising from the unique structure of pristine graphene, the electrochemical signal was found to be relatively small, precluding its use for a useful electrochemical sensor [38]; a more advantageous approach would be to use graphene that has been pre-treated to induce more edge plane like-defects across the surface of its basal plane. Interestingly this was shown to be the case using chemically-modified graphenes which contain different defect densities and varied amounts of oxygen-containing groups [39, 40]. It was demonstrated that differences in surface functionalities, structure and defects of various modified graphenes largely influence their electrochemical behaviour in detecting the oxidation of adenine, with later work demonstrating that few-layer graphene exhibits improved electroanalytical behaviour over that of single-layer graphene, multi-layered graphene (viz graphite), EPPG and unmodified GC electrode alternatives

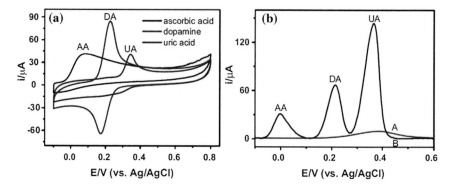

Fig. 4.10 **a** Cyclic voltammograms of 1.0 mM ascorbic acid (AA), 1.0 mM dopamine (DA) and 1.0 mM uric acid (UA) in 0.10 M PBS (pH 6.0) at NG modified GC electrode at a scan rate of 100 mVs^{-1}. **b** Differential pulse voltammograms for 1.0 mM *AA*, 0.05 mM DA and 0.10 mM *UA* in a 0.1 M PBS (pH 6.0) at a bare GC electrode (**a**) and a NG/GC electrode (**b**), respectively. Reprinted from Ref. [43] with permission from Elsevier

[39, 40]. Importantly, this work shows how it is possible to effectively tailor graphene to exhibit a range of favourable/required characteristics/properties for beneficial implementation and this will no doubt have a profound influence on the construction of graphene based sensors.

In other pursuits, the utilisation/fabrication of doped graphene structures (or the fabrication of novel three-dimensional hybrid/composite graphene materials) has been reported to be useful due to the significantly altered electrochemical properties, where modification of the graphene can result in improved conduction or electronic properties (DOS), increased disorder and/or edge plane accessibility [41, 42].

Numerous studies concerning the implementation of such graphene-based materials have been reported, for example towards the ultrasensitive analytical detection of cocaine [44], hydrogen peroxide [45], oxalate [46], ethanol [47], dopamine [48], nitric oxide [49], and various heavy metals (cadmium, lead, copper and mercury) [50]. One example, utilising nitrogen doped graphene (NG) towards the simultaneous determination of AA, DA and UA is shown in Fig. 4.10, where a high electro-catalytic activity towards the oxidation of all three analytes is reported. The NG electrochemical sensor displayed a wide linear response and low detection limits for AA, DA and UA, attributed by the authors to be due to its unique structure and properties originating from nitrogen doping [43]. Such work demonstrates that NG is a promising candidate for use as an advanced electrode material in electrochemical sensing and other electro-catalytic applications. It is important to note however that in the above work, although control experiments were performed with the unmodified GC electrode (which was used as the underlying support), comparisons were not made with other similar graphitic/carbon type materials that had undergone the same treatment (and of course to pristine graphene). Such control measures are crucial when determining the true origin of improved electrochemical performances, and thus it is important that they be extended to include cases where nano-composites are utilised (such as a metal nano-particle decorated graphene, or carbon-graphene based hybrid materials) in which case responses at each of the individual components must be considered (as well as with the addition of alternative, yet similar, graphitic structures); Appendix C provides a useful (non-exhaustive) summary of the various control experiments that should be considered by the graphene experimentalist.

4.2 Graphene Utilised in Energy Storage and Generation

There is an obvious need to significantly improve energy production and storage technologies to the point that fossil fuels can be adequately substituted. There have been substantial developments towards achieving this significant goal, but it is an ever moving target as there are major increases in demand in terms of the world energy consumption per annum [51]. One area of nanotechnology that is being extensively explored in order to meet the global energy challenge is the use of electrochemical applications utilising graphene and related structures. Graphene continues to receive extensive interest in electrochemical applications due to its reported large electric conductivity, large surface area, low production costs and reported electro-catalytic activity. [8, 52–54] We next explore the use of graphene in energy storage and conversion devices.

4.2.1 Graphene Supercapacitors

Supercapacitors find extensive utilisation within consumer electronics as their power intensity is higher than that of batteries since there are no chemical reactions during the charging and discharging processes. Instead energy storage is based upon electrochemical double-layer capacitance (EDLC) where energy is released by nanoscopic charge separation at the solid|liquid interface. Energy stored is inversely proportional to the double-layer thickness and such capacitors have high energy densities in comparison to conventional dielectric capacitors [51, 56]. Due to the absence of any chemical reactions, EDLC's can be rapidly charged and discharged, meaning that they can be used in hybrid vehicles such as in start-up devices and are useful in conjugation with fuel cells (for example in electric vehicles), where they can help extend the working lifetime of the fuel cells utilised.

Prior to the use of graphene, a carbon-based material such as activated carbon was extensively used as an electrode material in the construction of supercapac-itors due to its large surface area and low cost. However, in such structures, there are a lot of carbon atoms that cannot be accessed by the electrolyte ions, as illustrated in Fig. 4.11, and are effectively wasted; consequently this is a major factor that limits the specific capacitance (which is F/g) of activated carbon electrodes. Additionally, it is reported that the low electrical conductivity of

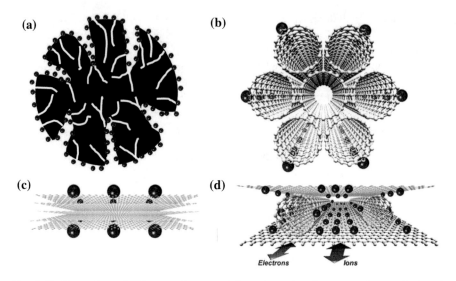

Fig. 4.11 Comparison of different carbon materials as electrodes of supercapacitors. **a** Activated carbon. Activated carbon has a high surface area. However, many of the micropores cannot be accessed by electrolyte ions. **b** Single-walled carbon nanotube (SWCNT) bundles. SWCNTs usually form bundles, limiting their surface area. Only the outmost surface can be accessed by electrolyte ions. **c** Pristine graphene. Graphene nano-sheets are likely to agglomerate through van der Waals interactions during the drying process. It would be difficult for electrolyte ions to access the ultra-small pores, especially for larger ions such as an organic electrolyte or at a high charging rate. **d** Graphene/CNT composite. SWCNTs can serve as a spacer between the graphene nano-sheets to give rise to rapid diffusion pathways for the electrolyte ions. Moreover, they can enhance electrical conduction for the electrons. The CNTs also serve as a binder to hold the graphene nano-sheets together preventing disintegration of the graphene structure into the electrolyte. Reproduced from Ref. [55] with permission from The Royal Society of Chemistry

activated carbon is also limiting its applications in high power density supercapacitors and results in a low specific capacitance per area of material. Following on from activated carbon, the use of carbon nanotubes (CNTs) as high-power electrode materials has been explored due to their reported higher electrical conductivity, enhanced charge transfer channels and high readily accessible surface area [51, 55]. However, as illustrated in Fig. 4.11, SWCNTs usually stack/aggregate in bundles and consequently only the outermost portion of CNTs can function for ion absorption, where the inner carbon atoms are not utilised, thus leading to low specific capacitance at CNT-based supercapacitors.

Table 4.2 compares the possible range of carbon based electrodes that can be employed along with key performance characteristics. The performance of a supercapacitor is mainly evaluated on the basis of possessing the following criteria: (i) a power density substantially greater than batteries with acceptably high energy densities (>10 Wh Kg^{-1}); (ii) an excellent cycle ability (more than 100 times that of batteries); (iii) fast charge-discharge processes (within seconds); (iv) low self-discharging; (v) safe operation; and (vi) low cost.

Table 4.2 A comparison of various carbon electrode materials for supercapacitors

Carbon	Specific surface area/m² g⁻¹	Density/g cm⁻³	Electrical conductivity/S cm⁻¹	Cost	Aqueous electrolyte		Organic electrolyte	
					$F\,g^{-1}$	$F\,cm^{-3}$	$F\,g^{-1}$	$F\,cm^{-3}$
Fullerene	1,100–1,400	1.72	10^{-8}–10^{-14}	Medium				
CNTs	120–500	0.6	10^{4}–10^{5}	High	50–100	<60	<60	<30
Graphene	2,630	>1	10^{6}	High	100–205	>100–205	80–110	>80–110
Graphite	10	2.26	10^{4}	Low				
ACs	1,000–3,500	0.4–0.7	0.1–1	Low	<200	<80	<100	<50
Templated porous carbon	500–3,000	0.5–1	0.3–10	High	120–350	<200	60–140	<100
Functionalized porous carbon	300–2,200	0.5–0.9	>300	Medium	150–300	<180	100–150	<90
Activated carbon fibers	1,000–3,000	0.3–0.8	5–10	Medium	120–370	<150	80–200	<120
Carbon aerogels	400–1,000	0.5–0.7	1–10	Low	100–125	<80	<80	<40

Reproduced from Ref. [57] with permission from The Royal Society of Chemistry

Shown within Table 4.2 is the possible performance of graphene as a supercapacitor. Graphene and chemically modified graphene sheets possess a high electrical conductivity, high surface area and outstanding mechanical properties that are comparable with, or even better than, CNTs [51, 55]. Of particular usefulness in energy storage and generation is the exceptionally high specific surface area of graphene, which is reported to be as large as 2675 m^2/g [56]; which is larger than that of activated carbon and CNTs, that are usually used in the ED-LCs—suggesting that graphene is a promising material for supercapacitors. Additionally, a graphene-based structure of individual sheets does not depend on the distribution of pores in the solid support to provide its large surface area, but instead, every chemically modified graphene sheet can "move" physically to adjust to different types of electrolytes (which is aided by graphene's reported high flexibility). As a result, access of the electrolyte to a very large surface area in graphene-based materials can be maintained while preserving the overall high electrical conductivity in the network [55].

However, there are also issues with pristine graphene derived supercapacitors since graphene agglomerates or restacks back to form graphite through the high cohesive van der Waals interactions during the drying process applied in obtaining graphene. As such, if the graphene sheets are stacked together it is difficult for ions to gain access to the inner layers in order to form the electrochemical double layers. In this case, the ions can only accumulate on the top and the bottom surfaces of the graphene sheets, leading to a lower specific capacitance since the stacked material cannot be fully used, as illustrated schematically in Fig. 4.11c. The hope of using graphene still exists, because, if used optimally, theoretically graphene is capable of generating a specific capacitance of 500 F g^{-1} (providing the entire 2,675 m^2 g^{-1}) is utilised; something which researchers aim to do, yet seldom achieve.

Cheng and co-workers demonstrated that a graphene/SWCNT composite is the best approach to the above problem [55]. They fabricated an electrode as shown in Fig. 4.11d, where a highly conductive spacer (SWCNTs) is utilised and as a result reduces the internal electrical resistance of the electrode and improves the power performance through preventing agglomeration between the graphene sheets—improving the accessibility for electrolyte ions [55]. Using this approach supercapacitors with high energy densities have been reported, with specific capacitances of ~290.6 and 201.0 F g^{-1} being obtained for a single electrode in aqueous and organic electrolytes respectively [55]. In the organic electrolyte the energy density was reported to reach 62.8 Wh kg^{-1} and the power density was 58.5 kW kg^{-1}. The addition of SWCNTs raised the energy density by 23 % and power density by 31 % when compared to the performance of 'just graphene' electrodes. The graphene/CNT electrodes exhibited an ultra-high energy density of 155.6 Wh kg^{-1} in an ionic liquid at room temperature, where additionally the specific capacitance increased by 29 % after 1,000 cycles (indicating the excellent cyclic-ability of their fabricated capacitor).

Other notable work [58] has used the approach shown in Fig. 4.12 where a specific capacitance value of 326 F g^{-1} at 20 mV s^{-1} was observed, which compared to 83 F g^{-1} where CNTs were not used.

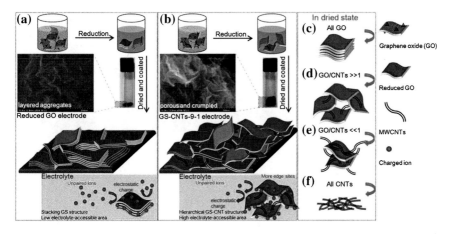

Fig. 4.12 Scheme for preparing **a** reduced GO electrode and **b** graphene sheet (GS)-CNTs-9-1 electrode. Note the schematic models of GS-CNT composites with various GS/CNTs ratios; where (**c**) GO are dispersed in the solution and GS aggregates/stacks are formed after reduction **d** GO and CNTs coexist in the solution and CNTs act as nanospacers to increase the interlayer space between GS after reduction, avoiding the aggregation issue **e** excess CNTs attach on the surface of GS resulting in a low exposure of surface area, and **f** CNTs are dispersed in the solution and CNTs aggregates are formed in the dry state. Note that GS refers to a graphene sheet. Reproduced from Ref. [58] with permission from The Royal Society of Chemistry

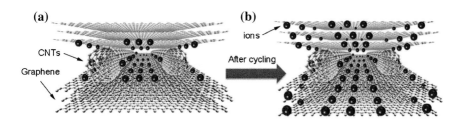

Fig. 4.13 Schematic illustrating the electro-activation to increase the electrode surface area after cycling. **a** Graphene sheets are likely to aggregate to form few-layered graphene which cannot be fully accessed by electrolyte ions at the first several charging and discharging cycles. **b** After a long time cycling, aggregated graphene sheets are separated by intercalated ions so there is a greater surface area available for electrolyte ions to result in an increase of specific capacitance after cycling. Reproduced from Ref. [55] with permission from The Royal Society of Chemistry

Furthermore the power density was reported as 70.29 kW kg^{-1}, highlighting the advantages of utilising such a fabricated electrode given that inspection with the literature reveals that such a response is highly competitive [58]. Cheng et al. [55] have also reported graphene and SWCNT composites exhibiting capacitances of $\sim$290.6 and 201.0 F g^{-1} in aqueous and organic electrolytes respectively. Figure 4.13 shows a schematic representation of the composite where the SWCNTs act as spacers to allow full access to the graphene surface. In the organic

Fig. 4.14 a Schematic of the fabrication process of a graphene-cellulose (GCP) membrane and **b** a photograph of a *GCP* membrane demonstrating its flexibility. Reproduced from Ref. [59] with permission from Wiley

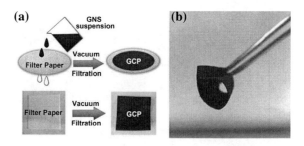

electrolyte the energy density reached 62.8 Wh kg^{-1} and the power density reached 58.5 kW kg^{-1} [55]. In this case the addition of SWCNTs raised the energy density by 23 % and the power density by 31 % over that of the graphene (with no CNT spacers) electrodes. The graphene/CNT electrodes exhibited an ultra-high energy density of 155.6 Wh kg^{-1} in an ionic liquid [55].

Other reports where novel graphene based supercapacitors have be readily and easily fabricated are freestanding and binder-less flexible supercapacitors, [59] consisting of three-dimensional (3D) interweaved structures comprising graphene nanosheets (GNS) and cellulose fibres (from the filter paper, see below); which exhibit exceptional mechanical flexibility and good specific capacitance. Figures 4.14 and 4.15 show that the method of fabrication is extremely simple, where the chosen GNSs are placed into an aqueous suspension and are then filtered through filter paper, as shown in Fig. 4.14, to produce a graphene-cellulose membrane (GCP).

These unique flexible supercapacitors have been shown to exhibit high stability and lose only 6 % of their electrical conductivity after being physically manipulated (bent) 1,000 times [59]. The high capacitance per geometric area is $\sim$81 mF cm^{-2} which is equivalent to 120 F g^{-1}, of which >99 % of the capacitance is retained over 5,000 cycles.

As mentioned in earlier chapters, graphene can be fabricated on a large scale via the reduction of GO, using either chemical, thermal or electron beam reduction [60–63]. Thermal reduction is reported to be able to give rise to few-layer graphene structures with less agglomeration, a higher specific surface area and higher electrical conductivity than graphenes produced via other fabrication routes [64, 65]. The effect of changing the thermal reduction temperature over the range, 200–900 °C has been explored with respect to the interlayer spacing, oxygen content, specific surface area and degree of disorder all being diligently monitored [66]. It was found that the highest capacity was 260 F g^{-1} at a charge/discharge current density of 0.4 A g^{-1} at a 200 °C thermal reduction temperature [66]. Interestingly, such specific capacitances are similar to that reported above where SWCNTs spacers were used (see Ref. [55]) suggesting that if the work of Ref. [66] is extended to include this, then even greater improvements can possibly be obtained.

However, it has recently been suggested that in certain cases, thermally reduced GO is no more beneficial, in specific selected energy applications, than that of amorphous carbon [67]. Additionally, note that the terminology used within the

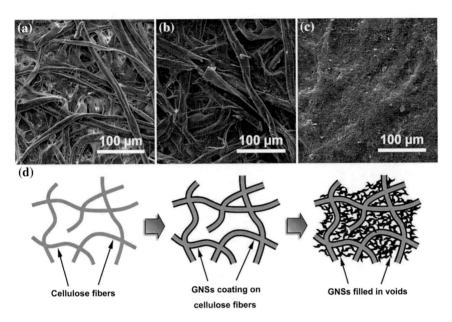

Fig. 4.15 SEM images of the filter paper or GCP surfaces with different GNS loading amounts. **a** 0 wt% (pristine filter paper), **b** 2.3 wt%, and **c** 7.5 wt%. **d** Illustration of the structural evolution of GCP as the graphene nanosheet loading increases. Reproduced from Ref. [59] with permission from Wiley

field is not consistent. For example, in the work of Zhao et al. [66] the thermal reduction of GO is reported via heating at a rate of 5 °C/min in a nitrogen atmosphere (where the temperature was varied from 200 to 900 °C for 2 h), whereas other work reports thermal exfoliation by heating at the desired temperature for 5 min in an air atmosphere [68]. Clearly both methods will result in entirely different graphene materials with the terminology ('reduced GO' or 'thermally reduced graphene') not really alerting the reader to these preparative differences.

Hybrid supercapacitors involving various components (as an alternative to the carbon-carbon based hybrid materials covered in detail above) are another avenue of research where graphene is beneficially utilised with pseudo-capacitive materials. Table 4.3 provides an overview of recent literature reports where it is clear that the synergic effect of the two capacitative methods are beneficial, revealing high capacitance values. Of note, solution-exfoliated graphene nanosheets (~ 5 nm thickness) have been coated onto 3D porous textile support structures (for high loading of active electrode materials and to facilitate the access of electrolytes to those materials) [69]. Figure 4.16 depicts the elegant fabrication approach. The hybrid graphene/MnO_2 based textile material yielded high-capacitance performances with specific capacitance up to 315 F g^{-1}. The ability to produce low cost, lightweight and potentially wearable energy storage devices is truly exciting.

Table 4.3 An overview of the specific capacitance and power output values of a range of graphene based materials and various other comparable materials for use as super-capacitors

Electrode material	Performance parameter		Cyclic ability	Comments	References
	Specific capacitance /F g^{-1}	Power density / kW kg^{-1}			
CNT/PANI	780	NT	After 1,000 cycles the capacitance decreased by 67 %	Obtained from CV measurement at a SR of 1 mV s^{-1}	[70]
GNS	150	NT	Specific capacitance was maintained with the specific current of 0.1 A g^{-1} for 500 cycles of charge/discharge	N/A	[71]
GNS	38.9	2.5	NT	Synthesised using a screen-printing approach and ultrasonic spray pyrolysis. Data obtained from CV measurement at a SR of 50 mV s^{-1}	[72]
GNS-Cobalt (II) hydroxide nano-composite	972.5	NT	NT	N/A	[73]
GNS/CNT/PANI	1,035	NT	After 1,000 cycles the capacitance decreased by only 6 % of the initial	Obtained from CV measurement at a SR of 1 mVs^{-1}	[70]
GNS-Nickel foam	164	NT	Specific capacitance remains 61 % of the maximum capacitance after 700 cycles	Obtained from CV measurement at a SR of 10 mV s^{-1}	[74]
GNS-SnO$_2$	42.7	3.9	NT	Synthesised using a screen-printing approach and ultrasonic spray pyrolysis. Data obtained from CV measurement at a SR of 50 mV s^{-1}	[72]

(continued)

Table 4.3 (continued)

Electrode material	Performance parameter		Cyclic ability	Comments	References
	Specific capacitance /F g⁻¹	Power density / kW kg⁻¹			
GNS-ZnO	61.7	4.8	NT	Synthesised using a screen-printing approach and ultrasonic spray pyrolysis. Data obtained from CV measurement at a SR of 50 mV s⁻¹	[72]
Graphene	205	10 @ 28.5 W h kg⁻¹	~90 % specific capacitance remaining after 1,200 cycles	Graphene was prepared from graphite oxide	[75]
MWCNT/PANI	463	NT	NT	Synthesised using in situ polymerisation. Capacitance obtained from CV measurement at a SR of 1 mV s⁻¹	[76]
Nickel (II) hydroxide nano-crystals deposited on GNS	1,335	NT	NT	Obtained at a charge/discharge density of 2.8 A g⁻¹	[77]
PANI	115	NT	NT	Synthesised using in situ polymerisation. Capacitance obtained from CV measurement at a SR of 1 mV s⁻¹	[76]
PANI/GOS	531	NT	NT	Nanocomposite with a mass ratio of PANI/graphene, 100:1. Capacitance obtained by charge-discharge analysis	[78]
RuO₂/GNS	570	10 @ 20.1 W h kg⁻¹	~97.9 % specific capacitance remaining after 1,000 cycles	N/A	[79]
MnO₂/Graphene	113.5	198 KW Kg⁻¹	NT	Only 2.7 % reduction observed for 1,000 cycles	

(continued)

Table 4.3 (continued)

Electrode material	Performance parameter		Cyclic ability	Comments	References
	Specific capacitance /F g⁻¹	Power density / kW kg⁻¹			
GO/MnO$_2$	328	25.8 KW Kg⁻¹			
Ni(OH)/Graphene	1,335			Charge/discharge at 2.8 A g⁻¹	[77]
GS	326.5	78.29 KW Kg⁻¹		CNTs used to stop graphene stacking	[58]
fGO	276	NT		Solvothermal method	[80]
RuO$_2$/Graphene	570	10 KW Kg⁻¹	97.9 % after 1,000 cycles		[79]
PSS/PDDA/GS	263	NT	90 % remaining following 1,000 cycles		[81]
Fe$_3$O$_4$/rGO	480	5506 W Kg⁻¹		Obtained at a discharge current density of 5 A g⁻¹	[82]
MnO$_2$/Graphene	113.5	198 KW Kg⁻¹	Only a 2.7 % reduction observed following 1,000 cycles		[83]
rGO/MnO$_2$	328	25.8 KW Kg⁻¹	NT		[84]
GO/MnO$_2$	310	NT	15,000 cycles with only a 4.6 % reduction		[85]
TEGO	230	NT	55 % capacitance retention reported.	Effect of thermal reduction temperature studied. 200 °C found optimal	[68]
TrGO	260.5	NT	NT		[66]

Key: *CNT* Carbon nanotube; *CV* Cyclic voltammetry; *fGO* Functionalised graphene oxide; *GNS* Graphene nano-sheet; *GOS* Graphene oxide sheets; *MWCNT* Multi-walled carbon nanotube; *N/A* Not applicable; *NT* Not tested; *PANI* Polyaniline; *PDDA* Poly(diallyldimethylamonium); *PSS* Poly(sodium 4-styrenesulfonate); *RuO$_2$* Hydrous ruthenium oxide; *SR* Scan rate; *TEGO* Thermal exfoliation of graphene; *TrGO* Thermally reduced graphene oxide

microfibers in textile   graphene nanosheets   graphene/MnO$_2$-textile
                                    -coated textile

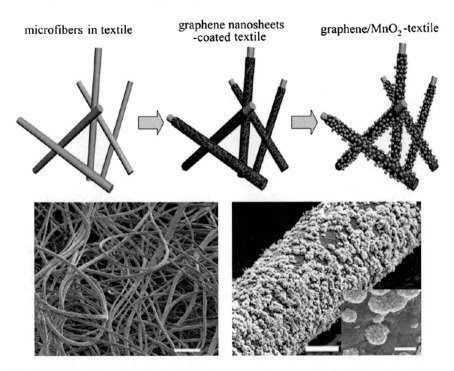

Fig. 4.16 Schematic illustrations of two key steps for preparing hybrid graphene/MnO$_2$-nanostructured textiles as high-performance electrochemical capacitor electrodes. *Left* to *right*: conformal coating of solution-exfoliated graphene nanosheets (*grey colour*) onto textile fibres, controlled electro-deposition of MnO$_2$ nanoparticles (*yellow dots*) on graphene-wrapped textile fibres. Accompanied by SEM images of a sheet of graphene-coated textile after 60 min of MnO$_2$ electro-deposition showing large-scale, uniform deposition of MnO$_2$ nanomaterials achieved on almost the entire fabric surface. Scale bar: 200 μm (*left*) and an SEM image of a typical micro-fibre with conformal coating of MnO$_2$ nanostructures (*right*). (Inset) high magnification SEM image showing the nano-flower structure of electrodeposited MnO$_2$ particles and a clear interface between MnO$_2$ nano-flower and underneath graphene nanosheets. Scale bars are 5 and 1 μm for main figure and inset respectively. Reprinted with permission from Ref. [69]. Copyright 2011 American Chemical Society

Following the theme of using 3D porous networks, since they permit large loading of active electrode materials and easy access of electrolytes to the electrodes, researchers have turned to fabricating 3D graphene foams [86].

Figure 4.17 depicts an example of such a 3D graphene foam, which are generally fabricated using a nickel skeleton upon which graphene is grown via CVD. The underlying skeleton is then etched away to leave a free-standing graphene structure [87]. This unique graphene structure has been demonstrated to give rise to high quality graphene with outstanding electrical conductivity, superior to macroscopic graphene structures from chemically derived graphene sheets [87]. Based on the ratio of the integrated intensity of the G and 2D bands obtained via Raman analysis, it was concluded that the walls of the foam are composed of

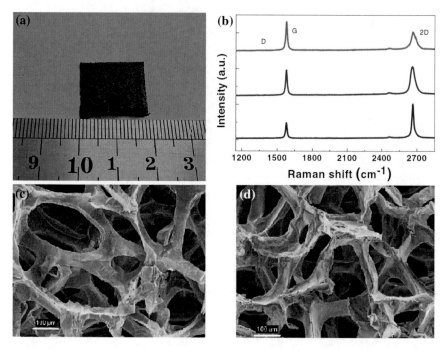

Fig. 4.17 **a** Photograph of a macroscopic 3D graphene foam network. The foam is flexible and can be easily handled and manipulated. **b** Raman spectra taken from several locations on the graphene foam clearly showing the Raman G band and 2D band peaks. **c** SEM image of the micro-porous graphene foam. **d** Teflon coated graphene foam showing similar pore structure and dimensions to the original graphene foam. Reproduced from Ref. [87] with permission from Wiley

mono- to few-layered graphene sheets; clearly such foam is better termed *quasi*-graphene 'as coined in Ref. [10]'. Note the absence of a defect related D band in the Raman spectrum indicates the high quality of graphene sheets prepared using the CVD process. Figure 4.17c shows SEM images of the micro-porous graphene foam in which an average pore size of ~ 200 µm is observed, while Fig. 4.17d shows a Teflon coated graphene foam which also exhibits a similar pore structure (and dimensions) as the original graphene foam [87]. The thickness of the Teflon coating is estimated to be ~ 200 nm (estimation performed by Teflon coating on to a flat surface which was subsequently cleaved for cross-sectional SEM analysis) [87]. Note that the authors were interested in obtaining a super-hydrophobic structure; a contact angle of 129.95° was observed at the 3D graphene foam, which increased to 150.21° following Teflon treatment; a value greater than 150° is termed a super-hydrophobic material.

In the specific case of energy storage, these unique 3D *quasi*-graphene foams have been widely explored. Figure 4.18 shows a 3D graphene structure that has been modified with cobalt oxide, giving rise to a specific capacitance of 1,100 F g^{-1} at a current density of 10 A g^{-1} [88]; the 3D structure can be readily modified

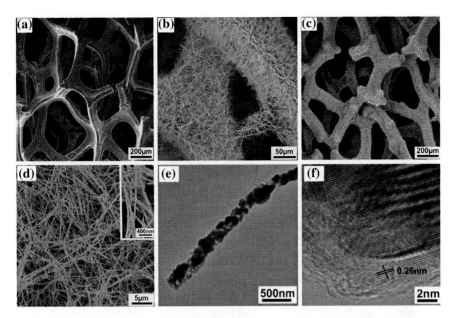

Fig. 4.18 SEM images of **a** 3D graphene foam **b** 3D graphene/Co$_3$O$_4$ nanowire composite. **c,** **d** Low- and high-magnification SEM images of graphene/Co$_3$O$_4$ nanowire composite. *Inset panel* **d** shows an enlarged view. **e, f** Low- and high-resolution TEM images of Co$_3$O$_4$ nanowire grown on the surface of 3D graphene foam. Reprinted with permission from Ref. [88]. Copyright 2012 American Chemical Society

with other metals and we expect this approach to be readily adopted by researchers. For example, as shown in Fig. 4.19, the 3D *quasi*-graphene foam has been modified with zinc oxide via the in situ precipitation of ZnO nanorods under hydrothermal conditions [89]. It was reported that the graphene/ZnO hybrids displayed superior capacitive performance than the unmodified alternative, with a high specific capacitance of ~ 400 F g^{-1} as well as an excellent cycle life; making these unique materials suitable for high-performance energy storage applications.

It is clear from this section that graphene is being beneficially utilised in the fabrication of supercapacitors and its incorporation into such devices is giving rise to some excellent specific capacitance and power densities. Other interesting work has addressed the graphene itself, that is, nitrogen doped graphene (produced via a plasma process) [90]. The doped graphene was shown to exhibit a specific capacitance of 280 F g^{-1}, 4 times greater than un-doped pristine graphene; the nitrogen-doping is reported to manipulate the electronic structure which improves the device performance [90]. As material scientists design and fabricate novel graphene structures, it is clear that significant enhancements in supercapacitors performance are likely to arise.

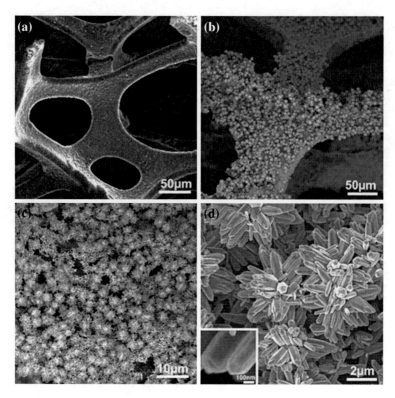

Fig. 4.19 SEM images of the samples. **a** SEM image of 3D graphene foam. **b–d** SEM images of graphene/ZnO hybrids with different magnifications. The inset of (**d**) shows the morphology of a single ZnO nanorod at a high magnification. Reproduced from Ref. [89] with permission from The Royal Society of Chemistry

4.2.2 Graphene Based Batteries/Li-Ion Storage

Lithium ion (Li-ion) based rechargeable batteries are a further class of energy storage devices where graphene has been employed due to its reported superior attributes. Li-ion based rechargeable batteries have been used extensively within portable electronics and are attractive for such purposes as they possess the ability to store and supply electricity over a long period of time. The electrode materials (i.e. anode/cathode) utilised in the fabrication of Li-ion batteries play a dominant role in the obtained performance, although each component of the battery is essential to its performance capabilities [54, 91]. Generally, a Li-ion device is composed of an anode, electrolyte and a cathode, as depicted in Fig. 4.20. On charging, the Li-ions are extracted from the cathode material and pass through the electrolyte until they reach and insert into the anode material; discharge reverses the procedure [51, 92]. Thus given that the output/performance of a battery is directly related to the Li-ion insertion/extraction process and efficiency at the

Fig. 4.20 A schematic illustration of a rechargeable lithium battery composed of cathode, anode, and electrolyte. Reproduced from Ref. [54] with permission from The Royal Society of Chemistry

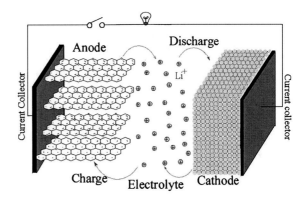

anode and cathode, the materials utilised for the two electrodes are fundamental to the battery's performance [54].

Currently the most commonly employed anode material for lithium based batteries is graphite, due to its high Coulombic efficiency (the ratio of the extracted Li to the inserted Li) [54] where it can be reversibly charged and discharged under intercalation potentials with a reasonable specific capacity [93]. However, researchers are looking for improvements in battery performance and wish to increase the relatively low theoretical capacity associated with graphite batteries (372 mA h g^{-1}) and the long diffusion distances of the Li-ions in such devices [51, 54].

Graphene has already shown itself experimentally as a beneficial replacement (anode material) for a new generation of Li-ion batteries due to it exhibiting higher specific capacities than many other electrode materials (including graphite); and additionally there are numerous theoretical papers having emerged to support this proposed ideology [93–96]. It is suggested that graphene's two-dimensional edge plane sites are expected to aid Li-ion adsorption and diffusion, allowing for reductions in charging times and an increased power output. An overview of various graphene based electrode materials that have been reported in the literature for use as a lithium based battery are listed and compared to other electrode materials, namely graphite and CNTs, in Table 4.4.

As reported in Sect. 4.2.1, the same problems experienced with capacitor based devices hamper Li-ion energy storage devices, in that aggregation of the graphene can limit applications and is an issue that needs to be overcome. Again, one solution (with respect to the fabrication of an all-carbon based hybrid electrode) is the use of CNTs as spacers between the graphene layers. For example elegant work has shown that a specific capacitance of 540 mA h g^{-1} can be obtained utilising a GNS electrode (a larger value than that of graphite, see above), which increases up to either 730 or 784 mA h g^{-1} through the incorporation of CNT or C_{60} macromolecules into the GNS structure respectively [97]. Again it is clear that the improved accessibility of graphene's surface area in such hybrid materials

Table 4.4 An overview of the specific capacitance and cyclic stabilities of a range of graphene based materials and various other comparable materials for the application of graphene as a Lithium-ion battery electrode

Compound	Specific capacitance /mA h g^{-1}	Cyclic stability	Comments	References
GNS	540	300 after 30 cycles at 50 mA g^{-1}	N/A	[93, 97]
GNS	1,264	848 after 40 cycles at 100 mA g^{-1}	GNSs in coin-type cells versus metallic lithium	[95]
GNS	1,233	502 after 30 cycles at the current density of 0.2 mA cm^{-2}	GNSs were prepared from artificial graphite by oxidation, rapid expansion and ultrasonic treatment	[65]
GNS/Fe$_3$O$_4$	1,026	580 after 100 cycles at 700 mA g^{-1}	N/A	[91]
GNS/SnO$_2$	860	570 after 30 cycles at 50 mA g^{-1}	GNSs were homogeneously distributed between the loosely packed SnO$_2$ nanoparticles in such a way that a nanoporous structure with a large amount of void spaces could be prepared	[93]
GNS/SnO$_2$	840	590 after 50 cycles at the current density of 400 mA g^{-1}	The optimum molar ratio of SnO$_2$/graphene was 3.2:1	[98]
GNS/with C$_{60}$ spacer molecules	784	NT	N/A	[97]
GNS/with CNT spacer	730	NT	N/A	[97]
Graphite	372	240 after 30 cycles at 50 mA g^{-1}	N/A	[93, 97]
Mn$_3$O$_4$/RGO	~900	~730 after 40 cycles at 400 mA g^{-1}	N/A	[99]
Ox-GNSs	~1400	Cyclic stability within the range of 800	Capacity loss per cycle of ~3 % for early cycles; this decreasing for subsequent cycles	[100]

Key: *C$_{60}$* Carbon 60; *CNT* Carbon nanotube; *GNS* Graphene nanosheet; *N/A* Not applicable; *NT* Not tested; *Ox* Oxidised; *RGO* Reduced GO

leads to enhanced capacitance capabilities with regards to Li-ion insertion, furthermore, nano-sized holes they may reside within the hybrid graphene structure have linked to high rate discharge capabilities in Li-ion batteries [101]. For example, work by Kung et al. which aimed to overcome the restrictions on performance caused by the low accessible volume (which is usually only a fraction of the physical volume) of stacked graphene layers resulted in the fabrication of 'holey graphene papers' (through facile microscopic engineering, i.e. controlled generation of in-plane porosity via a mechanical cavitation-chemical oxidation approach) that possessed abundant ion binding sites and thus exhibited enhanced ion diffusion kinetics and excellent high-rate Li-ion storage capabilities [102]. However, note that capacitance is not the only issue when looking at the performance of batteries, as discharge rates and cyclic abilities also need to be considered.

Other notable work was performed by Bai et al. [103] who explored the development of high quality graphene sheets that had been synthesised through an efficient oxidation process, followed by rapid thermal expansion and reduction by H_2. The group suggest that the number of graphene layers can be controlled by tuning the oxidation degree of GOs, with a higher oxidation degree resulting in fewer graphene layers. Various techniques were employed to determine the number of graphene layers including thermo-gravimetric analysis, SEM, AFM, TEM and Fourier Transform Infrared (FTIR) spectroscopy. It was determined through experimental analysis that the obtained graphene sheets with single, triple and quintuplicate layers as anode materials exhibit high reversible capacities of 1,175, 1,007 and 842 mA h g^{-1}, respectively. Such findings would strongly suggest that a higher reversible capacity is offered by those graphene sheets consisting of fewer layers. Furthermore, the cyclability offered was found to be highly encouraging given that after 20 cycles the graphene anode still maintained specific capacitance values of 8,46, 730 and 628 mA h g^{-1} for the single, triple and quintuplicate graphene layers respectively, that is, about 70 % retention of the reversible capacity [103]. Clearly such findings can potentially serve as the foundations on which further investigations involving the use of graphene with different numbers of layers can be facilitated [103].

In terms of utilising functionalised graphene within Li-ion batteries, Bhardwaj and co-workers [100] have reported the electrochemical Li-ion intake capacity of carbonaceous one-dimensional GNSs (obtained by unzipping pristine MWCNTs). The authors showed that oxidised-GNSs (ox-GNSs) outperformed the unmodified GNSs and also the original MWCNTs in terms of their energy density, obtaining a first charge capacity of $\sim$1,400 mA h g^{-1} (ox-GNS); with a low Coulombic efficiency for the first cycle, $\sim$53 and $\sim$95 % for subsequent cycles, which was shown to be notably superior to MWCNTs and GNSs and similar to that of graphite [100]. The cyclic capacity of the ox-GNSs was within the range of 800 mA h g^{-1}, with an early capacity loss per cycle of $\sim$3 % (this value decreasing upon subsequent cycles) [100]. Note that the cyclability is the capacity loss per charge/discharge

Fig. 4.21 **a** Schematic representation for the synthesis and structure of SnO_2 on graphene nanosheets (GNS). **b** Cyclic performances for (**a**) bare SnO2 nanoparticle (**b**) graphite (**c**) GNS, and (**d**) SnO2/GNS. Reprinted with permission from Ref. [93]. Copyright 2009 American Chemical Society

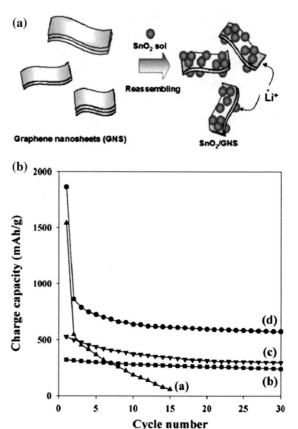

cycle, and is a useful parameter to evaluate battery performance over time, where in this case, MWCNTs and GNSs perform better than the proposed ox-GNSs with capacity losses of ~ 1.4 and ~ 2.6 % respectively, however, due to the higher initial capacitance of the ox-GNSs the performance of this material can still be regarded as superior due to the established cyclic capacity (stabilised capacitance value after continued cycling) remaining high, though improvements in the cyclability of the material would be desirable [100].

Hybrid graphene composite materials (other than carbon/graphene materials that have been highlighted above) have also been explored. In one example Paek et al. [93] demonstrated that a graphene/SnO_2 based nano-porous electrode exhibited a higher reversible capacity when compared to bare SnO_2, bare graphene, and bare graphite electrodes, and in addition to this, the graphene/SnO_2 electrode exhibited an improved cyclic performance when compared to the same electrodes [93]. Note here the efficient use of control experiments by the authors (however, an additional step would have been to utilise graphite for the fabrication of the hybrid material to ensure that observed improvements do in fact originate from the incorporation of graphene). The graphene/SnO_2 based nano-porous electrode was shown to exhibit a reversible capacity of 810 mA h g^{-1} [93]. Furthermore, its cyclic performance was drastically enhanced in comparison to bare SnO_2 nanoparticles, where after 30 cycles the charge capacity of the graphene/SnO_2 based electrode remained at 570 mA h g^{-1} (70 % retention), whereas the bare SnO_2 nanoparticles first charge capacity was 550 mA h g^{-1} which dropped rapidly to 60 mA h g^{-1} after only 15 cycles at 50 mA g^{-1}. Paek et al. [93] also reported superior cyclic performances over graphite modified nanoparticles, where the initial capacitance was $\sim$500 mA h g^{-1}, which dropped slightly during cycling—the bare graphene electrode resides only slightly above this level: Fig. 4.21 demonstrates the capacities and cyclic abilities of the mentioned composites along with a schematic representation of the fabricated graphene hybrid material [93].

Another useful metal oxide based graphene hybrid materials that have proven beneficial include the fabrication of a Mn_3O_4–graphene composite, which was reported to be a suitable high-capacity anode material for utilisation in Li-ion batteries [99]. Using a two-step solution-phase reaction to form Mn_3O_4 nanoparticles on reduced graphene oxide (rGO) sheets, one report has demonstrated that a Mn_3O_4–graphene anode exhibits a high specific capacity of $\sim$900 mA h g^{-1}, near the theoretical capacity of Mn_3O_4 ($\sim$936 mA h g^{-1}), with a good rate capability and cycling stability (a capacity of $\sim$730 mA h g^{-1} at 400 mA g^{-1} was retained after 40 cycles) [99]. The beneficial response was reported to arise from the intimate interactions between the graphene and the Mn_3O_4 nanoparticles. Note that even at a high current density of 1,600 mA g^{-1} the specific capacity was found to correspond to $\sim$390 mA h g^{-1}, which is higher than the theoretical capacity of graphite (see earlier). The authors performed control experiments utilising the synthesis of free (in the absence of graphene) Mn_3O_4 nanoparticles by the same process, where the performance was much worse; at a low current density of 40 mA g^{-1} the free Mn_3O_4 nanoparticles exhibited a capacity lower than 300 mA h g^{-1}, which further decreased to $\sim$115 mA h g^{-1} after only 10 cycles.

Whilst on the theme of metal oxide based hybrid materials, note that Fe_3O_4 exhibits great potential for use as an anode material with high capacity, low cost, eco-friendliness and natural abundance [91]. However because of a problem with rapid capacity fading during cycling, it has attracted the attention of scientists for use in the creation of hybrid materials [91]. A well-organised, flexible interleaved composite of GNSs decorated with Fe_3O_4 particles (through in situ reduction of iron hydroxide between GNSs) has been fabricated with the aim of overcoming the above issue; where the interleaved network of GNSs produce a pathway for electron transport as shown in Fig. 4.22. The GNS/Fe_3O_4 composite was reported to exhibit a reversible specific capacity approaching 1,026 mA h g^{-1} after 30 cycles (at 35 mA g^{-1}) and 580 mA h g^{-1} after 100 cycles (at 700 mA g^{-1}), as well as improved cyclic stability and an excellent rate capability [91]. In comparison, the capacities of the bare Fe_2O_3 and commercially available Fe_2O_4

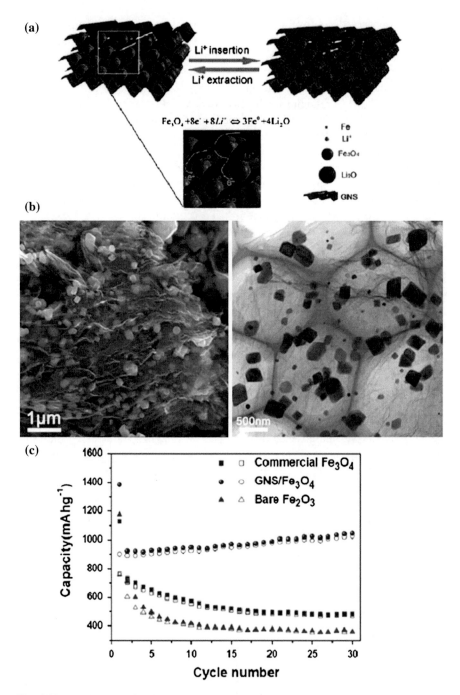

Fig. 4.22 a Schematic of a flexible interleaved structure consisting of graphene nanosheets (GNS) and Fe$_3$O$_4$ particles. **b (a)** SEM image of the cross-section of GNS/Fe$_3$O$_4$ composite, and **(b)** TEM. **c** Cycling performance of commercial Fe$_3$O$_4$ particles, GNS/Fe$_3$O$_4$ composite, and bare Fe$_2$O$_3$ particles at a current density of 35 mA g^{-1}—solid symbols, discharge; hollow symbols, charge. Reprinted with permission from Ref. [91]. Copyright 2010 American Chemical Society

particles following 30 cycles at 35 mA g^{-1} were found to decrease from 770 and 760 mA h g^{-1} to 475 and 359 mA h g^{-1} respectively, thus demonstrating poor cyclic abilities (the cyclic abilities are highlighted in Fig. 4.22) [91]. The multifunctional features of the GNS/Fe$_3$O$_4$ composite were reported to be as follows: (i) GNSs play a "flexible confinement" function to enwrap Fe$_3$O$_4$ particles, which can compensate for the volume change of Fe$_3$O$_4$ and prevent the detachment and agglomeration of pulverized Fe$_3$O$_4$, thus extending the cycling life of the electrode; (ii) GNSs provide a large contact surface for individual dispersion of well-adhered Fe$_3$O$_4$ particles and act as an excellent conductive agent to provide a highway for electron transport, improving the accessible capacity; (iii) Fe$_3$O$_4$ particles separate GNSs and prevent their restacking thus improving the adsorption and immersion of electrolyte on the surface of electro-active material; and (iv) the porosity formed by lateral GNSs and Fe$_3$O$_4$ particles facilitates ion transportation [91]. As a result, this unique laterally confined GNS/Fe$_3$O$_4$ composite can dramatically improve the cycling stability and the rate capability of Fe$_3$O$_4$ as an anode material for lithium ion batteries [91].

The fabrication of graphene based composite materials clearly precludes the issues surrounding the use of graphene for Li-ion batteries, resulting in greatly enhanced Li-ion insertion capacities and excellent cyclic performance in the majority of cases [51]. Alternatively, one can use other unique derivatives of graphene to alter graphene's properties with the aim of solving the aforementioned problems; such as doped graphene structures, for example nitrogen, potassium or boron-doped graphene, which have been utilised beneficially in energy technologies. In one such example Wu et al. [104] demonstrated that nitrogen- or boron-doped graphene (*N*- and *B*-doped respectively) can be used as a promising anode material for high-power and high-energy Li-ion batteries under high-rate charge/discharge conditions. The doped graphene was reported to show a high reversible capacity of >1,040 mA h g^{-1} at a low rate of 50 mA g^{-1}. Critically, the system was found to allow for rapid charge/discharge times of between 1 h to several tens of seconds, together with high-rate capability and excellent long-term cyclic ability. For instance, a very high capacity of ∼199 and 235 mA h g^{-1} was obtained for the *N*-doped graphene and *B*-doped graphene (at 25 A g^{-1}), respectively; requiring only ∼30 seconds to fully charge [104]. It was suggested that the unique 2D structure, disordered surface morphology, hetero-atomic defects, enhanced electrode/electrolyte wettability, increased inter sheet distance, improved electrical conductivity, and thermal stability of the doped graphenes are beneficial for rapid surface Li$^+$ absorption and ultrafast Li$^+$ diffusion and electron transport, thus making the doped materials superior to those of pristine chemically derived graphene and other carbonaceous materials [104].

It is apparent that current research utilising graphene as a Li-ion storage device indicates it to be beneficial over graphite based electrodes, exhibiting improved cyclic performances and higher capacitance for applications within Li-ion batteries, particularly when utilised as the basis of a hybrid/composite material and when utilising modified graphene structures.

4.2.3 Energy Generation

In contrast to the above sub-sections which consider energy storage, graphene has been reported to be beneficially utilised in the production of energy, such as in the use of fuel cells. There are many assortments of fuel cells, however they all work in the same general manner, converting chemical energy from a fuel into electricity, and they generally consist of the same components (an anode, a cathode and an electrolyte) [51]. Fuel cells are mainly classified by the type of electrolyte they use, however, with respect to the utilisation of graphene to enhance the given performance of specific fuel cells, the main requirements of a suitable catalyst support for favourable use in fuel cells are as follows: a high surface area, to obtain high metal dispersion; suitable porosity, to potentially boost gas flow/electrode accessibility; a high electrical conductivity; a widely applicable electro-catalytic activity; a high long-term stability under fuel cell operational conditions; and low production costs [105]. Table 4.5 provides a non-exhaustive overview of the various fuel cell devices that have been fabricated through the incorporation of graphene based materials.

In proton exchange membrane fuel cells (PEMFCs) platinum (Pt) based electro-catalysts are still widely utilised as anode and cathode electro-catalysts for hydrogen oxidation and for oxygen reduction reactions (ORRs) respectively. Work by Jafri et al. [108] utilised GNSs and nitrogen doped-GNSs as catalyst supports for Pt nanoparticles towards the ORR in PEMFCs, where the constructed fuel cells exhibited the power densities of 440 and 390 mW cm^{-2} for nitrogen doped-GNS-Pt and GNS-Pt respectively [108]. It is clear that the nitrogen doped device exhibited an enhanced performance, with improvements attributed to the process of nitrogen doping which created pyrrolic nitrogen defects that acted as anchoring sites for the deposition of Pt nanoparticles, and is also likely due to increased electrical conductivity and/or improved carbon-catalyst binding [108]. However, in contrast, earlier work by Zhang et al. [110] demonstrated that the fabrication of low-cost graphite sub-micron-particles (GSP) can be employed as a possible support for polymer electrolyte membrane (PEM) fuel cells, where Pt nanoparticles were deposited on GSP in addition to carbon black and CNT alternatives via an ethylene glycol reduction method. It was demonstrated that the Pt/GSP exhibited the highest electro-catalytic activity towards the ORR, and a durability study indicated the Pt/GSP was 2–3 times more durable than the CNT and carbon black alternatives [110]. It is evident however, that a graphene alternative should also have been incorporated into the control experiments in this work. Note that other work has reported that nitrogen-doped graphene can act as a metal-free electrode with a greatly enhanced electro-catalytic activity, long-term operation stability, and tolerance to crossover effect than Pt alternatives for the ORR via a four-electron pathway in alkaline solutions, producing water as a product [111].

Table 4.5 An overview of the power and current densities obtained through the utilisation of a range of graphene based materials and various other comparable materials as an electrode material in fuel cells

Electrode complex	Fuel/oxidant	Current/power density at maximum power	Power density versus time	Comments	References
Au/GNS/GOx	Glucose	156.6 ± 25 µA cm⁻² (load 15 kΩ)	24.3 ± 4 µW at 0.38 V (load 15 kΩ)	Tested daily with a 15 kΩ external load. Power output: 6.2 % loss after 24 h; 50 % loss after 7 days	[106]
Au/SWCNT/GOx	Glucose	86.8 ± 13 µA cm⁻²	7.8 ± 1.1 µW at 0.25 V (load 15 kΩ)	N/A	[106]
GMS-*E. coli*	Glucose	142 mW m⁻²	NT	N/A	[107]
GNS-Pt	H₂/O₂	390 mW cm⁻²	NT	PEMFC	[108]
GO-Pt	H₂/O₂	161 mW cm⁻²	NT	N/A	[109]
Nitrogen doped-GNS-Pt	H₂/O₂	440 mW cm⁻²	NT	PEMFC	[108]
Pt	H₂/O₂	96 mW cm⁻²	NT	N/A	[109]
SSM-*E. Coli*	Glucose	2668 mW m⁻²	NT	N/A	[107]

Key: *Au* Gold; *GMS* Graphene modified 'SSM'; *GNS* Graphene nanosheet; *GO* Graphene oxide; *GOx* Glucose oxidase; *N/A* Not applicable; *NT* Not tested; *PEMFC* Proton exchange membrane fuel cell; *Pt* Platinum; *SSM* Stainless steel mesh; *SWCNT* Single walled carbon nanotube

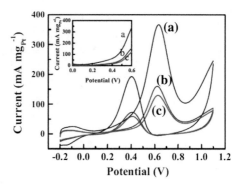

Fig. 4.23 The voltammetric responses of the oxidation of 0.5 M methanol in a 0.5 M H_2SO_4 aqueous solution at the Pt-Ni catalyst with a Pt/Ni molar ratio of 1:1 supported on graphene (*a*), SWNTs (*b*), and XC-72 carbon (*c*). The scan rate was 50 mV s^{-1}. The inset compares the starting potential for the oxidation of methanol at the Pt-Ni catalyst with a Pt/Ni molar ratio of 1:1 supported on graphene (*a*), SWNTs (*b*), and XC-72 carbon (*c*), respectively. Reproduced from Ref. [115] with permission from Elsevier

Direct methanol fuel cells have drawn great attention recently due to their high energy density, low pollutant emission, ease of handling (the liquid 'fuel'), and low operating temperatures (60–100 °C), however, the reported low electro-catalytic activity of present electrode materials towards methanol oxidation is hindering exploitation [112]. Graphene has been reported to enhance electro-catalytic activity of catalysts for fuel cell applications, in-particularly Xin et al. [112] have demonstrated that the utilisation of a platinum/GNS (Pt/GNS) catalyst revealed a high catalytic activity for both methanol oxidation and the ORR when compared to Pt supported on carbon black. Pt nanoparticles were deposited onto GNS via synchronous reduction of H_2PtCl_6 and GO suspensions using $NaBH_4$ with their results indicating that the current density of the Pt/GNS catalyst (182.6 mA mg^{-1}) outperforms the response of Pt/carbon black (77.9 mA mg^{-1}) towards the electrochemical oxidation of methanol. However, it was noted that prior heat treatment of the Pt/GNS catalyst improved the performance further by ∼3.5 times over that of the Pt/carbon black response; a response supported by previous literature. [112, 113] Moreover, work by Dong et al. [114] has demonstrated that Pt and Pt-Ru nanoparticles synthesised onto GNSs exhibit high electro-catalytic activity towards methanol and ethanol when contrasted against graphite alternatives, leading to a greatly reduced over-potentials and increased reversibility, thus these findings favour the routine use of graphene sheets as catalyst supports for both direct methanol and ethanol fuel cells [114].

Exceptionally fascinating work (exploring the oxidation of methanol for use within fuel cells) investigated the structure, composition and morphology of support materials, which were found to significantly affect the catalytic characteristics of Pt-based nanocatalysts [115]. Considering the effects of different carbon supports of graphene (produced via chemical reduction of GO), SWCNTs and Vulcan XC-72 carbon upon the electrocatalytic characteristics of the

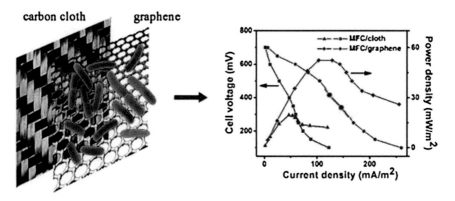

Fig. 4.24 A schematic representation of the fabricated anode consisting of bacteria growth onto a graphene sheet that has been electrochemically deposited onto a carbon cloth for use within MFCs (*left*), and the comparative output performances of the unmodified carbon cloth versus the graphene modified cloth (*right*). Reproduced from Ref. [118] with permission from Elsevier

nanocatalysts, as is evident in Fig. 4.23, the results demonstrate that the graphene-supported Pt-Ni catalyst has the highest electro-catalytic activity of the three carbon materials tested. This improved response at graphene was attributed to arise due to the oxygen-containing groups on the graphene surface, which can remove the poisoned intermediates and improve the electro-catalytic activity of the catalysts [115]. Further work considering the effects of graphene structure has been indicated that a vertically aligned, few-layer graphene electrode (with Pt nanoparticle deposits) exhibits a high resistance to carbon monoxide poisoning (and consequential deterioration in performance) when compared to commercially available alternatives [116].

As demonstrated above, GNSs are a good candidate for use as a supporting material in high-loading metal catalysts for potential applications in the fabrication of high-energy 'greener' solutions to current issues surrounding fuel cells; we expect the fundamental processes to be determined for metal immobilisation onto graphene, which is currently still not fully understood.

Other useful energy generation devices, which are not widely reported, are graphene derived Microbial Fuel Cells (MFCs) which offer a great opportunity to a obtain cleaner, more sustainable energy source [117]. Acting like bioreactors, MFCs generate electricity utilising the metabolic activity of microorganisms (bacteria) to decompose organic substances; such as their utilisation of waste products whilst efficiently generating electricity in order to meet increasing demands [117]. However, the practical applications of conventional MFCs are limited due to their relatively low power density and poor energy conversion efficiency, which results from sluggish electron transfer between the bacteria and the electrode [118].

Efforts are devoted to try and improve the electron transfer rate at the bacteria | electrode interface. For example Zhang et al. [107] has shown that the performance of a MFC based on *Escherichia coli* can be greatly enhanced (towards glucose) by utilising a graphene modified anode material. The authors modified a stainless steel mesh (SSM) with reduced graphene oxide and investigated its performance against a bare SSM and a polytetrafluoroethylene (PTFE) modified SSM (PMS), where the graphene modified SSM (GMS) exhibited an enhanced electrochemical performance and additionally a MFC equipped with a GMS anode delivered a maximum power density of 2,668 mW m^{-2} (which was 18 times larger than that obtained when using the bare SSM (142 mW m^{-2}) and 17 times larger than when utilising the PMS anode (159 mW m^{-2})) [107]. The enhanced performance of the GMS anode was attributed to graphene's high specific surface area and rough/flexible texture (as opposed to the SSM's and the PMS's relatively smooth surfaces) which promotes bacteria adhesion onto the anode surface and thus plays a key role in improving electron transfer and power output [107].

In other work Liu et al. demonstrated that an improved power density and energy conversion efficiency (of up to 2.7 and 3.0 times respectively, in comparison to the bare alternative) can be obtained by electrochemically depositing graphene onto a carbon cloth to fabricate an anode for a *Pseudomonas aeruginosa* mediator-less MFC [118]. A schematic representation of the fabricated electrode and the comparison of cell output are depicted in Fig. 4.24. Improvements in the response were attributed to be due to the high biocompatibility of graphene which promotes bacteria growth upon the electrode surface, resulting in the creation of more direct electron transfer activation centres and in this case stimulates excretion of mediating molecules for higher electron transfer rates to be realised [118]. In a further attempt to facilitate extracellular electron transfer in MFCs, work by Yuan et al. [119] has exploited a one-pot method to produce microbially-reduced graphene scaffolds. The authors added GO and acetate into a MFC in which GO is microbially reduced, leading to the in situ construction of a bacteria/graphene network on the anode surface [119]. Electrochemical measurements revealed that the number of exoelectrogens involved in extracellular electron transfer (EET) to the solid electrode increased due to the presence of graphene scaffolds, and the EET was facilitated in terms of electron transfer kinetics [119]. As a result, the maximum power density of the MFC was enhanced by 32 % (from 1,440 to 1,905 mW m^{-2}) and the Coulombic efficiency was improved by 80 % (from 30 to 54 %). This work demonstrates that the construction of the bacteria/graphene network is an effective method to improve the MFC performance.

It is clear from the above studies that low EET efficiency between the bacteria and the anode often limits the practical applications of MFCs and another critical challenge is to overcome the low bacterial loading capacities. Note that graphene is hydrophobic and in order to try and improve bacteria adhesion and biofilm formation the decoration of graphene surfaces with hydrophilic conducting polymers (such as the use of polyaniline (PANI)) has been attempted [120].

Last, enzymatic biofuel cells (EBFCs) possess the potential to be employed as an 'in-vivo' power source for implantable medical devices such as pacemakers [106]. The most striking feature of the EBFC is that they can utilise glucose or other carbohydrates copiously present in the human body as a fuel. However, EBFCs have major issues that need to be rectified, including low power densities and poor stability. There are very few reports of EBFCs utilising graphene, however, work is performed. One example by Liu et al. [106] concerned the use of GNSs within the construction of membraneless EBFCs. The authors used graphene to fabricate both the anode and cathode of a biofuel cell. The anode of the biofuel cell consisted of a gold electrode on which the authors co-immobilised graphene with glucose oxidase (GOx) using a silica sol-gel matrix while the cathode was constructed in the same manner except for the use of bilirubin oxidase (BOD) as the cathodic enzyme [106]. Voltammetric measurements were conducted to quantitatively evaluate the suitability and power output of employing a GNS as an electrode dopant and its performance was compared with a similar EBFC system constructed using SWCNTs. The maximum power density of the graphene based biofuel cell was shown to be 24.3 $\pm$ 4 μW at 0.38 V (load 15 kΩ), which is nearly two times greater than that of the SWCNTs EBFC (7.8 $\pm$ 1.1 μW at 0.25 V (load 15 kΩ)) and the maximum current density of the graphene based EBFC was found to be 156.6 $\pm$ 25 μA cm^{-2}, while for the SWCNT based EBFC is was 86.8 $\pm$ 13 μA cm^{-2}. To evaluate the stability of the graphene based EBFC the system was stored in pH 7.4 phosphate buffer solution at 4 °C and tested every day with a 15 kΩ external load, where after the first 24 h it had lost 6.2 % of its original power output. Later, the power output was found to decay slowly and became 50 % of its original power output after 7 days; which is substantially longer than other EBFC devices, and outperforms the SWCNT based EBFC [106]. The authors stated that the enhanced performance was based upon the larger surface area of graphene in comparison to SWCNTs, in addition to graphenes greater sp^2 character than other materials within its field (responsible for shuttling the electrons and assisting in the better performance of the EBFC) and the larger number of dislocations and electro-active functional groups present in graphene. Other work by Devadas and co-workers demonstrated the construction of a glucose/O$_2$ EBFC using an electrochemically reduced GO-MWCNT (ERGO-MWCNT) modified GC electrode as anode and a graphene-Pt composite modified GC electrode as a cathode [121]. The device achieved a maximum power density of 46 μW cm^{-2}, which, although this is lower than that obtained in the above work, compares well to a value of 57.8 μW cm^{-2} which was obtained at an alternative graphene based glucose/O$_2$ EBFC [122]. Although this is a relatively new area of graphene research, the reports highlighted above do show that graphene based composites are potential candidates for use in the development of efficient EBFCs of the future; we expect this field in-particular to be extended further.

4.3 Graphene: Final Thoughts

Throughout this chapter we have critically summarised the abundant literature concerning the use of graphene based materials in numerous electrochemical applications. We have aimed to provide useful insights into the core experimental issues facing graphene exploration. Resultantly, we have highlighted key works from which lessons can be learnt and the appropriate knowledge (and tools) gained, so that readers of this book are able to perform their own 'thorough' experiments and in doing so can contribute to the ever expanding graphene knowledgebase. Various 'key' experimental tips for the budding graphene experimentalist are listed in Appendix C.

In terms of the reported beneficial application of graphene and graphene based devices throughout the literature, we have allowed readers to make their own mind up on whether graphene has revolutionised these fields. One sentiment however is clear, although there are many outstanding properties of graphene which are due to the low proportion of defects across the graphene surface, nano-engineering of graphene based devices is often necessary to introduce structural confirmations, defects or specific impurities/composite materials that allow the desired functionality and electronic structures to give rise to useful electrochemical activity. As new routes to fabricate or modify graphene are produced and the array of graphene based structures available to experimentalists continues to evolve, in terms of electrochemistry, the next generation of graphene based devices will continue to emerge.

A report on 10th December 1991 concerned a debate on the future of buckyballs (C_{60}) in the House of Lords, initiated by Lord Errol of Hale who asked Her Majesty's Government: '*What steps they are taking to encourage the use of buckminsterfullerene in science and industry?* The debate rambled on for some time, amid growing confusion, until Baroness Seear felt it incumbent upon herself to ask, perplexedly, '*My Lords, forgive my ignorance, but can the noble lord say whether this thing is animal, vegetable or mineral?*' Lord Reay provided an eloquent explanation, describing the football-like shape and constitution of this molecule 'known to chemists as C_{60}'. His response, however, had Lord Renton baffled: '*My Lords, is it the shape of a rugger football or a soccer football?*', he enquired.

And so the debate went on, until at last Lord Campbell of Alloway ventured to ask that all-important question: '*My Lords, what does it do?*', to which Lord Reay responded '*My Lord, it is thought it may have several possible uses: for batteries; as a lubricant; or as a semi-conductor. All that is speculation. It may turn out to have no uses at all*'. At that, Earl Russell finally had the last word:

'*My Lords, can one say that it does nothing in particular and does it very well*'.
Source: Designs on C_{60}; Chemistry World, September issue, 1996.

References

1. A. Heller, B. Feldman, Chem. Rev. **108**, 2482–2505 (2008)
2. J.P. Metters, R.O. Kadara, C.E. Banks, Analyst **136**, 1067–1076 (2011)
3. Y. Wang, Y. Li, L. Tang, J. Lu, J. Li, Electrochem. Commun. **11**, 889–892 (2009)
4. W.-J. Lin, C.-S. Liao, J.-H. Jhang, Y.-C. Tsai, Electrochem. Commun. **11**, 2153–2156 (2009)
5. C.E. Banks, R.G. Compton, Analyst **130**, 1232–1239 (2005)
6. X. Kang, J. Wang, H. Wu, J. Liu, I.A. Aksay, Y. Lin, Talanta **81**, 754–759 (2010)
7. Y. Wang, D. Zhang, J. Wu, J. Electroanal. Chem. **664**, 111–116 (2012)
8. D.A.C. Brownson, C.E. Banks, Analyst **135**, 2768–2778 (2010)
9. Y. Shao, J. Wang, H. Wu, J. Liu, I.A. Aksay, Y. Lin, Electroanalysis **22**, 1027–1036 (2010)
10. D.A.C. Brownson, L.C.S. Figueiredo-Filho, X. Ji, M. Gomez-Mingot, J. Iniesta, O. Fatibello-Filho, D.K. Kampouris C.E. Banks, J. Mater. Chem. A **1**, 5962–5972 (2013)
11. S. Park, R.S. Ruoff, Nat. Nanotechnol. **4**, 217–224 (2009)
12. X. Ji, C.E. Banks, A. Crossley, R.G. Compton, ChemPhysChem **7**, 1337–1344 (2006)
13. R.L. McCreery, Chem. Rev. **108**, 2646–2687 (2008)
14. C.E. Banks, T.J. Davies, G.G. Wildgoose R.G. Compton, Chem. Commun. **7**, 829–841 (2005)
15. A. Chou, T. Böcking, N.K. Singh, J.J. Gooding, Chem. Commun. **7,** 842–844 (2005)
16. D.A.C. Brownson, C.W. Foster, C.E. Banks, Analyst **137**, 1815–1823 (2012)
17. J. Li, S. Guo, Y. Zhai, E. Wang, Anal. Chim. Acta **649**, 196–201 (2009)
18. Y.-R. Kim, S. Bong, Y.-J. Kang, Y. Yang, R.K. Mahajan, J.S. Kim, H. Kim, Biosens. Bioelectron. **25**, 2366–2369 (2010)
19. Y. Wang, Y. Wan, D. Zhang, Electrochem. Commun. **12**, 187–190 (2010)
20. J. Peng, C. Hou, X. Hu, Int. J. Electrochem. Sci. **7**, 1724–1733 (2012)
21. M.S. Goh, M. Pumera, Anal. Bioanal. Chem. **399**, 127–131 (2011)
22. D.A.C. Brownson, L.J. Munro, D.K. Kampouris, C.E. Banks, RSC Adv. **1**, 978–988 (2011)
23. B.R. Kozub, N.V. Rees, R.G. Compton, Sens. Actuators B **143**, 539–546 (2010)
24. D.A.C. Brownson, D.K. Kampouris, C.E. Banks, Chem. Soc. Rev. **41**, 6944–6976 (2012)
25. T.-W. Hui, K.-Y. Wong, K.-K. Shiu, Electroanalysis **8**, 597–601 (1996)
26. M.C. Granger, M. Witek, J. Xu, J. Wang, M. Hupert, A. Hanks, M.D. Koppang, J.E. Butler, G. Lucazeau, M. Mermoux, J.W. Strojek, G.M. Swain, Anal. Chem. **72**, 3793–3804 (2000)
27. D.A.C. Brownson, A.C. Lacombe, D.K. Kampouris, C.E. Banks, Analyst **137**, 420–423 (2012)
28. D.A.C. Brownson, C.E. Banks, Analyst **136**, 2084–2089 (2011)
29. M.S. Goh, M. Pumera, Anal. Chem. **82**, 8367–8370 (2010)
30. D.A.C. Brownson, M. Gomez-Mingot, C.E. Banks, Phys. Chem. Chem. Phys. **13**, 20284–20288 (2011)
31. D.A.C. Brownson, R.V. Gorbachev, S.J. Haigh, C.E. Banks, Analyst **137**, 833–839 (2012)
32. D.A.C. Brownson, C.E. Banks, Phys. Chem. Chem. Phys. **14**, 8264–8281 (2012)
33. C.X. Lim, H.Y. Hoh, P.K. Ang, K.P. Loh, Anal. Chem. **82**, 7387–7393 (2010)
34. S. Wu, X. Lan, F. Huang, Z. Luo, H. Ju, C. Meng, C. Duan, Biosens. Bioelectron. **32**, 293–296 (2012)
35. K.J. Huang, D.J. Niu, J.Y. Sun, C.H. Han, Z.W. Wu, Y.L. Li, X.Q. Xiong, Colloids Surf. B **82**, 543–549 (2011)
36. N.G. Shang, P. Papakonstantinou, M. McMullan, M. Chu, A. Stamboulis, A. Potenza, S.S. Dhesi, H. Marchetto, Adv. Funct. Mater. **18**, 3506–3514 (2008)
37. K.R. Ratinac, W. Yang, J.J. Gooding, P. Thordarson, F. Braet, Electroanalysis **23**, 803–826 (2011)
38. E.P. Randviir, C.E. Banks, RSC Adv. **2**, 5800–5805 (2012)
39. M.S. Goh, A. Bonanni, A. Ambrosi, Z. Sofer, M. Pumera, Analyst **136**, 4738–4744 (2011)
40. M.S. Goh, M. Pumera, Anal. Chim. Acta **711**, 29–31 (2012)

41. B. Guo, L. Fang, B. Zhang, J.R. Gong, Insciences J. **1**, 80–89 (2011)
42. X. Huang, X. Qi, F. Boey, H. Zhang, Chem. Soc. Rev. **41**, 666–686 (2012)
43. Z.-H. Sheng, X.-Q. Zheng, J.-Y. Xu, W.-J. Bao, F.-B. Wang, X.-H. Xia, Biosens. Bioelectron. **34**, 125–131 (2012)
44. B. Jiang, M. Wang, Y. Chen, J. Xie, Y. Xiang, Biosens. Bioelectron. **32**, 305–308 (2012)
45. Y. Huang, S.F.Y. Li, J. Electroanal. Chem. **690**, 8–12 (2013)
46. H. Li, J. Chen, S. Han, W. Niu, X. Liu, G. Xu, Talanta **79**, 165–170 (2009)
47. W.H. Gao, Y.S. Chen, J. Xi, S.Y. Lin, Y.W. Chen, Y.J. Lin, Z.G. Chen, Biosens. Bioelectron. **41**, 776–782 (2013)
48. X.L. Niu, W. Yang, H. Guo, J. Ren, J.Z. Gao, Biosens. Bioelectron. **41**, 225–231 (2013)
49. Y. Chen, B. Gao, J.X. Zhao, Q.H. Cai, H.G. Fu, J. Mol. Model. **18**, 2043–2054 (2012)
50. Y. Wei, C. Gao, F.-L. Meng, H.-H. Li, L. Wang, J.-H. Liu, X.-J. Huang, J. Phys. Chem. C **116**, 1034–1041 (2012)
51. D.A.C. Brownson, D.K. Kampouris, C.E. Banks, J. Power Sources **196**, 4873–4885 (2011)
52. A.K. Geim, K.S. Novoselov, Nat. Mater. **6**, 183–191 (2007)
53. D. Chen, L. Tang, J. Li, Chem. Soc. Rev. **39**, 3157–3180 (2010)
54. M. Liang, L. Zhi, J. Mater. Chem. **19**, 5871–5878 (2009)
55. Q. Cheng, J. Tang, J. Ma, H. Zhang, N. Shinya, L.-C. Qin, Phys. Chem. Chem. Phys. **13**, 17615–17624 (2011)
56. C. Liu, Z. Yu, D. Neff, A. Zhamu, B.Z. Jang, Nano Lett. **10**, 4863–4868 (2010)
57. L.L. Zhang, R. Zhou, X.S. Zhao, J. Mater. Chem. **20**, 5983–5992 (2010)
58. S.-Y. Yang, K.-H. Chang, H.-W. Tien, Y.-F. Lee, S.-M. Li, Y.-S. Wang, J.-Y. Wang, C.-C.M. Ma, C.-C. Hu, J. Mater. Chem. **21**, 2374–2380 (2011)
59. Z. Weng, Y. Su, D.-W. Wang, F. Li, J. Du, H.-M. Cheng, Adv. Energy Mater. **1**, 917–922 (2011)
60. H.-J. Shin, K.K. Kim, A. Benayad, S.M. Yoon, H.K. Park, I.-S. Yung, M.H. Jin, H.-K. Jeong, J.M. Kim, J.-Y. Choi, Y.H. Lee, Adv. Funct. Mater. **19**, 1987–1992 (2009)
61. M.J. McAllister, J.L. Li, D.H. Adamson, A.A. Abdala, J. Liu, M. Herra-Alonso, D.L. Milius, R. Car, R.K. Prud'homme, I.A. Aksay, Chem. Mater. **19**, 4396–4404 (2007)
62. H.C. Schniepp, J.L. Li, M.J. McAllister, H. Sai, M. Herrera-Alonso, D.H. Adamson, R.K. Prud'homme, R. Car, D.A. Saville, I.A. Aksay, J. Phys. Chem. B **110**, 8535–8539 (2006)
63. L. Chen, Z. Xu, J. Li, C. Min, L. Liu, X. Song, G. Chen, X. Meng, Mater. Lett. **65**, 1229–1230 (2011)
64. W.Z. Bao, F. Miao, Z. Chen, H. Zhang, W.Y. Jang, C. Dames, C.N. Lau, Nat. Nanotechnol. **4**, 562–566 (2009)
65. G. Peng, H. Song, X.H. Chen, Electrochem. Commun. **11**, 1320–1324 (2009)
66. B. Zhao, P. Liu, Y. Jiang, D. Pan, H. Tao, J. Song, T. Fang, W. Xu, J. Power Sources **198**, 423–427 (2012)
67. C.H.A. Wong, A. Ambrosi, M. Pumera, Nanoscale **4**, 4972–4977 (2012)
68. Q. Du, M. Zheng, L. Zhang, Y. Wang, J. Chen, L. Xue, W. Dai, G. Ji, J. Cao, Electrochim. Acta **55**, 3897 (2010)
69. G. Yu, L. Hu, M. Vosgueritchian, H. Wang, X. Xie, J.R. McDonough, X. Cuu, Y. Cui, Z. Bao, Nano Lett. **11**, 2905–2911 (2011)
70. J. Yan, T. Wei, Z. Fan, W. Qian, M. Zhang, X. Shen, F. Wei, J. Power Sources **195**, 3041–3045 (2010)
71. X. Du, P. Guo, H. Song, X. Chen, Electrochim. Acta **55**, 4812–4819 (2010)
72. T. Lu, Y. Zhang, H. Li, L. Pan, Y. Li, Z. Sun, Electrochim. Acta **55**, 4170–4173 (2010)
73. S. Chen, J. Zhu, X. Wang, J. Phys. Chem. C **114**, 11829–11834 (2010)
74. Y. Chen, X. Zhang, P. Yu, Y. Ma, J. Power Sources **195**, 3031–3035 (2010)
75. Y. Wang, Z. Shi, Y. Huang, Y. Ma, C. Wang, M. Chen, Y. Chen, J. Phys. Chem. C **113**, 13103–13107 (2009)
76. J. Yan, T. Wei, B. Shao, Z. Fan, W. Qian, M. Zhang, F. Wei, Carbon **48**, 487–493 (2010)
77. H. Wang, H.S. Casalongue, Y. Liang, H. Dai, J. Am. Chem. Soc. **132**, 7472–7477 (2010)
78. H. Wang, Q. Hao, X. Yang, L. Lu, X. Wang, Electrochem. Commun. **11**, 1158–1161 (2009)

79. Z.-S. Wu, D.-W. Wang, W. Ren, J. Zhao, G. Zhou, F. Li, H.-M. Cheng, Adv. Funct. Mater. **20**, 3595–3602 (2010)
80. Z. Lin, Y. Yai, O.J. Hildreth, Z. Li, K. Moon, C.-P. Wong, J. Phys. Chem. C **115**, 7120–7125 (2011)
81. Z. Li, J. Wang, X. Liu, J. Ou, S. Yang, J. Mater. Chem. **21**, 3397–3403 (2011)
82. W. Shi, J. Zhu, D.H. Sim, Y.Y. Tay, Z. Lu, X. Zhang, Y. Sharma, M. Srinivasan, H. Zhang, H.H. Hng, Q. Yan, J. Mater. Chem. **21**, 3422–3427 (2011)
83. Z. Fan, J. Yan, T. Wei, L. Zhi, G. Ning, T. Li, F. Wei, Adv. Funct. Mater. **21**, 2366–2375 (2011)
84. Q. Chen, J. Tang, J. Ma, H. Zhang, N. Shinya, L.-C. Quin, Carbon **49**, 2917–2925 (2011)
85. J. Yan, Z. Fan, T. Wei, W. Qian, M. Zhang, F. Wei, Carbon **48**, 3825–3833 (2010)
86. Z. Chen, W. Ren, L. Gao, B. Liu, S. Pei, H.-M. Cheng, Nat. Mater. **10**, 424–428 (2011)
87. E. Singh, Z. Chen, F. Houshmand, W. Ren, Y. Peles, H.-M. Cheng, N. Koratkar, Small **9**, 75–80 (2013)
88. X.-C. Dong, H. Xu, X.-W. Wang, Y.-X. Huang, M.B. Chan-Park, H. Zhang, L.-H. Wang, W. Huang, P. Chen, ACS Nano **6**, 3206–3213 (2012)
89. X. Dong, Y. Cao, J. Wang, M.B. Chan-Park, L. Wang, W. Huang, P. Chen, RSC Adv. **2**, 4364–4369 (2012)
90. H.M. Jeong, J.W. Lee, W.H. Shin, Y.J. Choi, H.J. Shin, J.K. Kang, J.W. Choi, Nano Lett. **11**, 2472–2477 (2011)
91. G. Zhou, D.-W. Wang, F. Li, L. Zhang, N. Li, Z.-S. Wu, L. Wen, G.Q. Lu, H.-M. Cheng, Chem. Mater. **22**, 5306–5313 (2010)
92. M. Winter, J.O. Besenhard, M.E. Spahr, P. Novak, Adv. Mater. **10**, 725–763 (1998)
93. S.-M. Paek, E. Yoo, I. Honma, Nano Lett. **9**, 72–75 (2009)
94. D. Pan, S. Wang, B. Zhao, M. Wu, H. Zhang, Y. Wang, Z. Jiao, Chem. Mater. **21**, 3136–3142 (2009)
95. P. Lian, X. Zhu, S. Liang, Z. Li, W. Yang, H. Wang, Electrochim. Acta **55**, 3909–3914 (2010)
96. C. Uthaisar, V. Barone, Nano Lett. **10**, 2838–2842 (2010)
97. E.J. Yoo, J. Kim, E. Hosono, H.-S. Zhou, T. Kudo, I. Honma, Nano Lett. **8**, 2277–2282 (2008)
98. X. Wang, X. Zhou, K. Yao, J. Zhang, Z. Liu, Carbon **49**, 133–139 (2011)
99. H. Wang, L.-F. Cui, Y. Yang, H.S. Casalongue, J.T. Robinson, Y. Liang, Y. Cui, H. Dai, J. Am. Chem. Soc. **132**, 13978–13980 (2010)
100. T. Bhardwaj, A. Antic, B. Pavan, V. Barone, B.D. Fahlman, J. Am. Chem. Soc. **132**, 12556–12558 (2010)
101. T. Takamura, K. Endo, L. Fu, Y. Wu, K.J. Lee, T. Matsumoto, Electrochim. Acta **53**, 1055–1061 (2007)
102. X. Zhao, C.M. Hayner, M.C. Kung, H.H. King, ACS Nano **5**, 8739–8749 (2011)
103. X. Tong, H. Wang, G. Wang, L. Wan, Z. Ren, J. Bai, J. Bai, J. Solid State Chem. **184**, 982–989 (2011)
104. Z.-S. Wu, W. Ren, L. Xu, F. Li, H.-M. Cheng, ACS Nano **5**, 5463–5471 (2011)
105. E. Antolini, Appl. Catal. B **123–124**, 52–68 (2012)
106. C. Liu, S. Alwarappan, Z. Chen, X. Kong, C.-Z. Li, Biosens. Bioelectron. **25**, 1829–1833 (2010)
107. Y. Zhang, G. Mo, X. Li, W. Zhang, J. Zhang, J. Ye, X. Huang, C. Yu, J. Power Sources **196**, 5402–5407 (2011)
108. R.I. Jafri, N. Rajalakshmi, S. Ramaprabhu, J. Mater. Chem. **20**, 7114–7117 (2010)
109. B. Seger, P.V. Kamat, J. Phys. Chem. C **113**, 7990–7995 (2009)
110. S. Zhang, Y. Shao, X. Li, Z. Nie, Y. Wang, J. Liu, G. Yin, Y. Lin, J. Power Sources **195**, 457–460 (2010)
111. L. Qu, Y. Liu, J.-B. Baek, L. Dai, ACS Nano **4**, 1321–1326 (2010)
112. Y. Xin, J.-G. Liu, Y. Zhou, W. Liu, J. Gao, Y. Xie, Y. Yin, Z. Zou, J. Power Sources **196**, 1012–1018 (2011)

113. Y.M. Li, L.H. Tang, J.H. Li, Electrochem. Commun. **11**, 846–849 (2009)
114. L. Dong, R.R.S. Gari, Z. Li, M.M. Craig, S. Hou, Carbon **48**, 781–787 (2010)
115. Y. Hu, P. Wu, Y. Yin, H. Zhang, C. Cai, Appl. Catal. B **111–112**, 208–217 (2012)
116. N. Soin, S.S. Roy, T.H. Lim, J.A.D. McLaughlin, Mater. Chem. Phys. **129**, 1051–1057 (2011)
117. D. Pant, G.V. Bogaert, L. Diels, K. Vanbroekhoven, Bioresour. Technol. **101**, 1533–1543 (2010)
118. J. Liu, Y. Qiao, C.X. Guo, S. Lim, H. Song, C.M. Li, Bioresour. Technol. **114**, 275–280 (2012)
119. Y. Yuan, S. Zhou, B. Zhao, L. Zhuang, Y. Wang, Bioresour. Technol. **116**, 453–458 (2012)
120. J. Hou, Z. Liu, P. Zhang, J. Power Sources **224**, 139–144 (2013)
121. B. Devadas, V. Mani, S.-M. Chen, Int. J. Electrochem. Sci. **7**, 8064–8075 (2012)
122. W. Zheng, H.Y. Zhao, J.X. Zhang, H.M. Zhou, X.X. Xu, Y.F. Zheng, Y.B. Wang, Y. Cheng, B.Z. Jang, Electrochem. Commun. **12**, 869–871 (2010)

Author Biographies: Received 3-March-14

Craig E. Banks is an Associate Professor / Reader in nano and electrochemical technology at The Manchester Metropolitan University, Manchester UK. Craig has published >260 papers ($h = 42$) in addition to 4 books, 14 book chapters and he is the inventor of 17 patent families. Craig has spun out 2 companies from his research. He was awarded the RSC Harrison–Meldola Memorial Prize (2011) for his 'contributions to the understanding of carbon materials, in particular graphene and its application as an electrode material'. Craig is also an Associate Editor of the RSC journal "Analytical Methods" and a Honorary Professor at Xiangtan (湘潭大學) University. His interests lie in the pursuit of studying the fundamental understanding and applications of nano-electrochemical systems such as graphene, carbon nanotube and nanoparticle derived sensors and developing novel electrochemical sensors *via* screen printing and related techniques. Additionally his research peruses energy storage in the form of graphene based supercapactiors and Li and Na ion batteries.

Dale A. C. Brownson is a Research Associate in graphene engineering and graphene electrochemistry at The Manchester Metropolitan University, Manchester UK. He has published >25 papers ($h = 12$) and has contributed 2 book chapters. Dale received the RSC Ronald Belcher Award (2013) for 'achievements towards the fundamental understanding of graphene as an electrode material and for contributions to the implications of this knowledge for the evolution of improved electroanalytical sensors'. Dale's work has focused on expanding the horizons of graphene electrochemistry, which encompasses the exploration of fundamental understanding in addition to applications in sensing and energy related devices. In addition to his fundamental interests regarding fabricating and investigating 'advanced' carbon nano-materials for beneficial implementation within electrochemical devices, his other research interests include the forensic applications of chemistry.

D. A. C. Brownson and C. E. Banks, *The Handbook of Graphene Electrochemistry*, DOI: 10.1007/978-1-4471-6428-9, © Springer-Verlag London Ltd. 2014

Appendix A
A Letter to the Nobel Committee

The letter to the Nobel Committee below was written in response to the 'Scientific Background' document issued by the Nobel Committee in support of the 2010 Nobel Prize in Physics being awarded jointly to Geim and Novoselov "*for groundbreaking experiments regarding the two-dimensional material graphene*" (an updated version of the 'Scientific Background' document can be found at Ref. [1]). The letter has been reproduced in full from Ref. [2] for scientific clarity.

Letter from Walt de Heer

November 17, 2010

To: The Nobel Committee,

Class for Physics of the Royal Swedish Academy of Sciences

The Nobel Prize in Physics is the most prestigious scientific achievement award and it is expected that the award be based on diligent and independent investigations. The scientific background document published by the Class for Physics of the Royal Swedish Academy of Sciences that accompanies the 2010 Nobel Prize in Physics is considered to reflect this process and it is therefore presumed to be accurate. I am recognized to be an authoritative source in the research area of the 2010 Nobel Prize in Physics. I can attest to the fact that this document contains serious inaccuracies and inconsistencies, so that the document presents a distorted picture that will be echoed in the community at large if the errors remain uncorrected. I list below several of the more serious errors with suggested changes.

1. Figure 3.3 is a reproduction of a figure in Novoselov et al.'s 2004 paper [1]. The figure caption incorrectly states the measurements were made on *graphene* (a single layer of carbon). The 2004 caption states that the measurement was performed on a FLG sample (i.e. ultrathin graphite composed of several graphene layers). In fact Noveoselov's 2004 paper

D. A. C. Brownson and C. E. Banks, *The Handbook of Graphene Electrochemistry*,
DOI: 10.1007/978-1-4471-6428-9, © Springer-Verlag London Ltd. 2014

does *not report any electronic transport measurements on graphene*. The band-structure figure accompanying this figure represents graphite and not graphene and the magnetoresistance measurements are explicitly graphitic. The Manchester group published graphene transport measurements in 2005 [2]. Please note also, that the right panel of Fig. 4.4 is incorrectly labelled and ambiguously credited.

2. Page 2 states: "*It should be mentioned that graphene-like structures were already known of in the 1960s but there were experimental difficulties [13–16] and there were doubts that this was practically possible.*" The references all relate to graphene under various conditions. None of the references discuss experimental difficulties nor do they express doubts about the practical possibility (to produce) graphene. For example the respected graphite scientist, H-P Boehm, who later coined the name "graphene", published his 1962 observations of graphene in a most highly regarded journal (Ref. [13]) and demonstrated beyond reasonable doubt the existence of freestanding graphene. He certainly showed that the existence of graphene was practically possible. The Nobel committee cites this work and then contradicts its main conclusion without explanation. Boehm's work has stood the test of time and has been reproduced by others. References [14–16] demonstrate that besides freestanding graphene, other forms of graphene are also practically possible. The document must explain how it arrives at the opposite conclusion or replace the sentence with, for example: "*It should be mentioned that graphene structures were already known of before 2004 [13–16]*".

3. Page 1 states: "*It was well known that graphite consists of hexagonal carbon sheets that are stacked on top of each other, but it was believed that a single such sheet could not be produced in isolated form. It, therefore, came as a surprise to the physics community when in 2004, Konstantin Novoselov, Andre Geim and their collaborators [1] showed that such a single layer could be isolated and that it was stable.*" This critically important assertion is repeated several times in the document without justification. In fact, the (chemical) stability of graphene did not come as a surprise, even for those who were unaware of Boehm's experiments. Despite Novoselov et al.'s claim in Ref. [1], the chemical stability of graphene did not violate any physical principle and its existence was not doubted in any research paper. Graphene had previously been observed and characterized as a two-dimensional crystal by several research groups [4]. Careful reading of Ref. [1] suggests that Novoselov et al. had confused highly stable covalently bonded two-dimensional macromolecules (like micron-sized graphene flakes), with chemically unstable freestanding two dimensional metal crystals, causing them to presume that theoretically graphene should also be chemically unstable. None of the references cited in Ref. [1] questions

the existence of graphene in any circumstance, contradicting the statement in the document that its observation 'came as a complete surprise'. On the contrary, several references cited in Ref. [1] actually show images of graphene under various conditions. Had graphene's existence in any form truly violated accepted physical principles, then its observation would have resulted in a flurry of activity to explain the discrepancy. In reality, Ref. [1] did not give rise to a single paper re-examining the chemical stability of isolated graphene.

The document must satisfactorily justify the controversial statement quoted above which certainly does not reflect the consensus opinion of experts in the field and it is overwhelmingly contradicted by facts as pointed out in item 2, above. The sentence might be replaced with "*It was well known that graphite consists of hexagonal carbon sheets that are stacked on top of each other and researchers were developing methods to deposit single sheets on substrates. In 2005, Konstantin Novoselov, Andre Geim and their collaborators demonstrated a simple method to deposit and to identify a single graphene sheet on an oxidized silicon carbide wafer [2]*" with a reference to their 2005 PNAS article [2], and not their 2004 Science article Ref. [1], as explained in item 5.

4. Page 7 states: *The mobility of graphene is very high which makes the material very interesting for electronic high frequency applications [37]. Recently it has become possible to fabricate large sheets of graphene. Using near-industrial methods, sheets with a width of 70 cm have been produced [38, 39].*

Geim and Novoselov's method obviously cannot be used for electronic applications; for such purposes, other, previously established graphene production methods are used. The large graphene sheets were made by a CVD method (first described in the 1990s) developed by Ruoff et al. The first actual high frequency transistors were made with epitaxial graphene on silicon carbide at Hughes Research Laboratories in 2009 and at IBM in 2010 using concepts and methods (first described in the 1970s) developed by de Heer et al. [3] Earlier in the document, epitaxial graphene is referred to as "carbon layers" on silicon carbide as if it were somehow different than graphene. Well before 2004, epitaxial graphene on silicon carbide had been described as a 2-dimensional crystal that is free floating above the substrate (cf. Ref. [15] of the document). It has been shown to exhibit every essential graphene property and photoemission measurements have become icons for graphene's band structure. De Heer's research preceded, and, most importantly, developed entirely independently from Geim and Novoselov's research. (In 2004 he performed the first graphene transport measurements: the incorrect thickness measurement in Ref. [3]a was corrected in Ref. [3]b.) The document gives the impression that de Heer's research on graphene based electronics (initiated in 2001) was contingent, stimulated or in some other way motivated by Geim and Novoselov. This is not the case, and the document should clarify this.

5. The Summary paragraph, page 7 states: *The development of this new material opens new exciting possibilities. It is the first crystalline 2D-material and it has unique properties, which makes it interesting both for fundamental science and for future applications. The breakthrough was done by Geim, Novoselov and their co-workers; it was their paper from 2004 which ignited the development. For this they are awarded the Nobel Prize in Physics 2010.*

Geim and Novoselov developed a very simple method to produce and observe microscopic graphene slivers on oxidized, degenerately doped silicon wafers. This method was copied by many and provides an ideal method to produce graphene samples for two-dimensional transport studies. The development of this experimental technique was very important for the field of mesoscopic physics, and as pointed out in the document, this was Geim and Novoselov's most important contribution.

However this method and its application to graphene by Novoselov et al. was *not* reported in 2004 [1] but in 2005 [2]. In Ref. [1] the ultrathin graphite flakes (FLG) whose transport properties were measured, were produced by a more cumbersome method that certainly would not have attracted so much attention [cf. supporting on-line material for Ref. [1]. In fact Ref. [1] does not report measurements nor characterization of graphene: instead, it presents evidence of a microscopic sliver of graphene protruding from an ultrathin graphitic flake, not unlike those observed earlier by others (i.e. Shioyama op. cit. Ref. [1]). It is relevant that Ref. [1] is often wrongly cited for "the discovery of graphene" and for the "Scotch tape method", even by the authors of Ref. [1]. This misrepresentation of Ref. [1] should be corrected in the document.

Further note that de facto isolated graphene had been identified and characterized as a 2D-crystalline material in many reports prior to 2004 (see for example [4] for a review). The characterization of graphene as a *new* 2D material is incorrect. This might be corrected in the document along the lines of the second paragraph in this item.

The authors of the Scientific Background document misquoted essential facts pertaining to Ref. [1]. An independent review of this document would be helpful to assure that the statements are clear, unambiguous, and factually correct.

We hope that the committee reviews these facts, corrects and publishes an erratum to the scientific background document so that it rises to the exacting standards expected of it.

Sincerely yours,

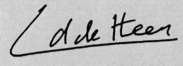

Walt de Heer
Regents Professor of Physics
Georgia Institute of Technology

1. K. Novoselov et al., Electric field effect in atomically thin carbon films. Science **306**, 666–669 (2004)
2. K. Novoselov et al., Two-dimensional atomic crystals. Proc. Nat. Acc. Sci. **102**, 10451 (2005)
3. a. C. Berger et al., Ultrathin epitaxial graphite: 2D electron gas properties and a route toward graphene-based nanoelectronics. J. Phys. Chem. **108**, 19912 (2004). b. W.A. de Heer, Epitaxial graphene. Sol. St. Comm. **43**, 92 (2007)
4. N.R. Gall et al., Two dimensional graphite films on metals and their intercalation. Int. J. Mod. Phys. B **11**, 1865 (1997)

Appendix B
Useful Concepts for Data Analysis

In analytical chemistry, and indeed electroanalysis, a calibration curve is produced so that analysts can determine how their system responds towards the target analyte under investigation. In-particular the calibration curve gives the electrochemist a benchmark of their electroanalytical system, such as a graphene modified electrode towards sensing analyte X.

Shown in Fig. B.1 is a typical calibration curve which is made by recording the signal output of a piece of apparatus after presenting different concentrations of the target analyte X to the system, which is electrochemical in this case. This is usually achieved by preparing a solution and running the response in absence of analyte X (termed "blank"), after which additions of analyte X are made into the solution to produce a final concentration of analyte X (carefully taking into account the dilution factor) from a concentrated stock solution. Additions are made in order to deduce the response (or sensitivity) of the system/electrode configuration. Experimentalists should note that calibrated micropipettes should be utilised along with high quality grade solutions with the stock solution having analyte X dissolved in the same solution as one is adding to, i.e. same buffer solution/composition to avoid any misinterpretations such as might be observed from simple pH changes.

As noted above Fig. B.1 shows a typical calibration curve, however it also highlights upon this calibration curve the dynamic linear range. The data (the increasing concentrations of the analyte and the instrument response) can be fit to a line of best fit (using the linear portion only) using linear regression analysis. This yields a model described by the equation $y = q_1 x + q_0$, where y is the electrochemical response, q_1 represents the sensitivity (gradient of the slope), and q_0 is a constant that describes the background/blank response which has a contribution from non-Faradaic processes.

When constructing a calibration plot, the approach should be repeated to obtain statistically meaningful data. Reporting the analytical performance based upon one calibration plot is unprofessional and not realistic of the electroanalytical system being developed; yet such practices still carry on within the literature. Note that the approach above will determine the *intra*-electrochemical response since the same electrode is being used, where ideally the *inter*-electrochemical response should also be explored. Useful definitions are as follows.

D. A. C. Brownson and C. E. Banks, *The Handbook of Graphene Electrochemistry*,
DOI: 10.1007/978-1-4471-6428-9, © Springer-Verlag London Ltd. 2014

Fig. B.1 Typical calibration plot highlighting the dynamic range and other pertinent analytically useful parameters that are used to benchmark (electro) analytical systems

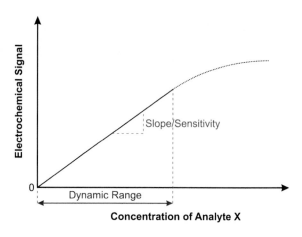

Intra-electrochemical response:

This is also known as repeatability i.e. the ability to repeat the same procedure with the same analyst, using the same reagent and equipment in a short interval of time, e.g. using the same electrode with measurements made within a day and obtaining similar results.

Inter-electrochemical response:

The ability to repeat the same method under different conditions, e.g. change of analyst, reagent, or equipment; or on subsequent occasions, e.g. change of electrode, measurements recorded over several weeks or months, this is covered by the 'between batch' precision or reproducibility, also known as inter-assay precision.

Note that in addition to the above, the electrode response (i.e. calibration plot) should be obtained for each of the following scenarios: (i) without polishing during the measurements, when using a solid electrode such as a GC electrode; (ii) polishing between measurements, again when using solid electrodes; and (iii) a new electrode for each measurement, as is possible when using SPEs.

B.1   Mean/Average and Standard Deviation

The sample mean, $\bar{x}$, is defined as the mean or average of a limited number of samples drawn from a population of experimental data, as defined by:

$$\bar{x} = \frac{\sum_i x_i}{N} \tag{B.1}$$

where $\bar{x}$ is defined as the value of an individual experimental value, $\sum_i x_i$ is the sum of all the experimental values, and N is the number of experimental values used. Standard

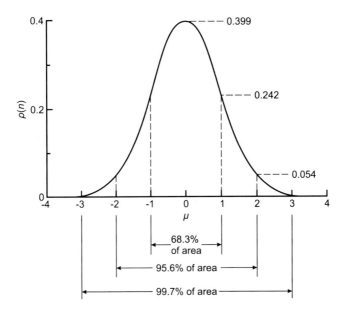

Fig. B.2 The normal curve of error: the Gaussian distribution plotted in units of the standard deviation σ with its origin at the mean value μ showing the percentage of values lying between $\mu - n\sigma$ and $\mu + n\sigma$ where $\rho(n)$ is the probability density function. In this example, the area under the curve is 1; half the area falls between $\mu = \pm 0.67$

deviation, s, is used as a measure of precision which describes how two or more numbers are in agreement if the exact same method or procedure is used and is given by:

$$s = \sqrt{\frac{\sum\limits_{i}^{N}(x_i - \bar{x})^2}{N-1}} \tag{B.2}$$

The above equation assumes the errors can be equally positive or negative relative to the mean value and if this is the case, there are no systematic errors, then x is equal to the mean of the normal or Gaussian distribution. The Gaussian distribution is shown in Fig. B.2 with mean, μ and the standard deviation, σ: 68.3 % of the measured values lie between $\mu - \sigma$ and $\mu + \sigma$, while 95.6 % lie between $\mu - 2\sigma$ and $\mu + 2\sigma$, while 99.7 % are between $\mu - 3\sigma$ and $\mu + 3\sigma$. This is the basis of the confidence internals and confidence limits.

B.2 Reproducibility

The reproducibility of a method is of prime interest to the analyst since this will give a better representation of the precision achieved during routine use as it includes the variability from a greater number of sources. The precision is a measure of random error and is defined as the agreement between replicate measurements of the same sample. It is expressed as the percentage coefficient of variance (%CV) or Percentage Relative Standard Deviation (%RSD) of the replicate measurements, defined as:

$$\%CV/\%RSD = (\text{standard deviation}/\text{mean}) \times 100 \tag{B.3}$$

B.3   Limit of Detection (LOD) and Limit of Quantification (LOQ)

As part of any analytical laboratory, internal quality control involves calibration and validation of working practices with the limit of detection and limit of quantification routinely performed. In terms of Fig. B.1, the detection limit is a useful analytical and electroanalytical parameter to benchmark the developed electrochemical system against other literature reports. Its definition however is not very straight forward. The IUPAC definition, as given in the *Gold Book*, is as follows [3]:

"The minimum single result which, with a stated probability, can be distinguished from a suitable blank value. The limit defines the point at which the analysis becomes possible and this may be different from the lower limit of the determinable analytical range. The limit of detection is expressed as the concentration, c_L, or the quantity, q_L, derived from the smallest measure, x_L, that can be detected with reasonable certainty for a given analytical procedure, is given by:

$$x_L = \bar{x}_{blank} + k s_{blank}$$

where $\bar{x}_{blank}$ is the mean of the blank value (analyte-free sample), s_{blank} is the standard deviation of the blank, and k is a numerical factor chosen according to the confidence (statistical) level desired."

Ideally, the definition of the LOD and LOQ need to be relatively simple if it is to be widely utilised. Unfortunately, this is not the case and there are considerable efforts from the IUPAC in order to clarify this [4–6].

A common and widely used approach, still in use, is based on the IUPAC and American Chemical Society definition, which uses the mean blank signal, $\bar{y}_B$ as the basis for the calculation of the signal LOD which is defined as:

$$LOD = (y_D - \bar{y}_B)/q_1 = K_D s_b/q_1 \tag{B.4}$$

where $\bar{y}_D$ is the signal value and $\bar{y}_B$ is the mean blank signal, K_D is the numerical factor chosen according to the confidence (statistical) level desired. In terms of the LOQ, the expression is:

$$LOQ = (y_Q - \bar{y}_B)/q_1 = K_Q s_b/q_1 \tag{B.5}$$

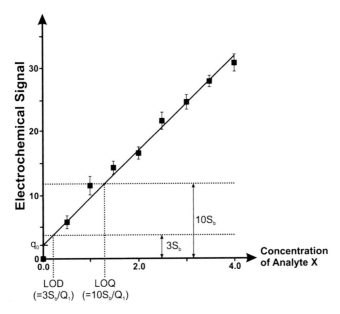

Fig. B.3 An illustration of the two alternatives of the standard approach, the LOD and LOQ calculation. Each calibration point represents an average of replicative measurements

where usually it is generally accepted that K_D is equal to 3 while K_Q is equal to 10; in academic papers where the LOD is reported using this approach, it is usually referred to as simply 3-sigma.

The above approaches work through the use of calibration plots (see Fig. B.3) where the measurement is performed enough times to be statistically relevant and where the error of the line of best fit is used with the relevant equation, such that, in the case of the LOD, $LOD = 3s_b/q_1$ can be readily deduced as shown in Fig. B.3. Note that there are limitations where over- and under- estimation of the LOD and LOQ can become possible using this approach and interested readers are directed towards Ref. [4, 5]. Indeed, if the calibration plot is of high quality, s_b will be very small, leading to an extremely low detection limit being reported. In practice, this will never be achieved due to the limitation in the experimental procedure of the instrumentation (such as noise and drift at low signal levels and so on) as well as chemical interferences, since typically the LOD and LOQ are reported in model systems (buffer solutions only).

Table B.1 Critical values for the rejection quotient, $Q_{critical}$

Number of observations/number of data, N	90 % confidence	95 % confidence
3	0.941	0.970
4	0.765	0.829
5	0.642	0.710
6	0.560	0.625
7	0.507	0.568
8	0.468	0.526
9	0.437	0.493
10	0.412	0.466
11	0.392	0.444
12	0.376	0.426
13	0.361	0.140
14	0.349	0.396
15	0.338	0.384
16	0.329	0.374
17	0.320	0.365
18	0.313	0.356
19	0.306	0.349
20	0.300	0.342

B.4 Evaluating the Quantify of Sets of Data

Q-test:

In the development of an electroanalytical protocol, comparisons need to be made with a well-established and accepted 'accurate' or 'gold standard' procedure. The two important statistical analysis methods to allow such comparisons are the t-test and F-test. Before the application of these, the Q-test needs to be performed, that is, the obtained data should be tested for potential outliers (anomalies), which are data values which appear to be unreasonably distant (or outlying) from the others comprising the data set, and to do this *objectively*, involves the implementation of the Q-test [5].

The Q-test allows the rejection of outliers from data by calculating, Q, defined by:

$$Q = \frac{(suspect\ value - nearest\ value)}{(\max.value - \min.value)} \quad (B.6)$$

From the experimentally obtained values, arrange the data in order of increasing values and deduce Q from Eq. (B.6), then compare this with tabulated $Q_{critical}$ values; Table B.1 displays critical values for confidence levels of 90 and 95 %.

If the calculated Q value exceeds that of $Q_{critical}$ then the suspect value is rejected from the data set. As such, the rejected value will either be the maximum or the minimum values; if a value is rejected then the Q-test should be repeated with the new smaller data set until no further data are removed. For further information and larger tables of critical values see Refs. [7, 8].

t-test:

The t-test allows the comparison of the average/mean of two data sets and compares the actual difference between two means in relation to the variation in the data (expressed as the standard deviation of the difference between the means). A good analogy for the t-test (also known as the student t-test) is as follows: if there was quality control of dim sum being produced by two separate dim sum chefs, a suitable hypothesis is that there are no differences in weight of the dim sum produced between the two chefs; the student's t-test[1] will allow one to deduce from experimental data, for example by weighing a sample of the dim sum, if the two data sets are consistent (in that the dim sum produced are essentially the same) or depart significantly from ones expectation (such that the produced dim sum are dramatically different and customers are either getting a great deal or a very bad deal). We next consider the t-test in terms of data that will likely be generated in the course of developing ones electrochemical protocol.

• *Case 1: Comparison between an accepted value and an experimental data set*

This type of scenario is used to compare an experimental mean with a value that is obtained from a sample, where the value is certified through analytical means known as a Certified Reference Material (CRM). CRMs are put through rigorous testing procedures to validate accurate concentration levels and therefore there is a high degree of confidence in these analytically determined concentrations. In order to compare an experimental value with a CRM value to validate a method/procedure, the following t-test is implemented.

Assuming a Gaussian distribution, such as that shown in Fig. B.2, the percentage of the distribution lying between specified limits around the mean value can be calculated; this is the width of the confidence interval corresponding to that percentage for an infinite population (infinite number of degrees of freedom). For a finite number of analyses, Eq. (B.7) describes the limits within which the true mean must lie at a given confidence level with respect of the experimental mean.

---

[1] W. S. Gosset, a Chemist (New College, Oxford) developed the t-test while working for the Guinness brewery (Dublin, Ireland) after developing and applying it to monitor the quality of stout. As Gosset was not allowed to publish the t-test by his employer, he used the pseudonym "Student", (publication: Student (1908a). *The probable error of a mean*. Biometrika, 6, 1–25 and Student (1908b). *Probable error of a correlation coefficient*. Biometrika 6, 302–310) and hence the student–t-test arises. Interested readers are directed to: Fisher Box, Joan (1987). "Guinness, Gosset, Fisher, and Small Samples". *Statistical Science* **2**(1): 45–52 for a full history.

Table B.2 The t-distribution; values of t for selected confidential intervals (CI)

Confidence interval significance level, P	90 % CI 0.10	95 % CI 0.05
Degrees of freedom		
1	6.314	12.706
2	2.920	4.303
3	2.353	3.182
4	2.132	2.776
5	2.015	2.571
6	1.943	2.447
7	1.895	2.365
8	1.860	2.306
9	1.833	2.262
10	1.812	2.228
∞	1.645	1.960

$$\mu = \bar{x} \pm \frac{ts}{\sqrt{N}} \tag{B.7}$$

In order for there to be no significant difference:

$$|\mu - \bar{x}| \geq \frac{ts}{\sqrt{N}} \tag{B.8}$$

We can re-arrange Eq. (B.8) for convenience giving rise to:

$$\pm t = (\bar{x} - \mu)\sqrt{\frac{N}{s}} \tag{B.9}$$

where μ is the value of the certified reference material, t is the student's t-value, obtained for $N-1$ degrees of freedom, at a pre-selected confidence interval (typically a 95 % confidence interval). The t-values are obtained from tables, such as that presented in Table B.2. Values of t are tabulated according to the number of degrees of freedom, in the case $(N-1)$ and depend on the number of analyses performed and the confidence interval, i.e what percentage of the hypothetical Gaussian distribution is to be included. Confidence levels are also presented as probabilities, P, of differences being found, e.g. a confidence level of 95 % corresponds to $P = 0.05$; an infinite number of degrees of freedom correspond exactly to the Gaussian distribution.

Case 1 can be used to compare whether or not the data for the given experimental method is considered reliable and valid in contrast to the CRM value. In the case of experimentally obtained (electrochemical) data, following the elimination of outliers using the Q-test (see above), the deduced mean value ($\bar{x}$) and standard deviation (s) along with the CRM value (μ) and the number of data used (N) to deduce the mean value allows one to determine a value for t. One then consults the t-table (such as that presented in Table B.2) at the desired confidence internal and compares this to the deduced t value. If the calculated t-value is lower than the tabulated t value at the confidence internal, there is no statistical difference; in this case the proposed/utilised electroanalytical method is considered a valid experimental procedure. Conversely, if the calculated t-value is higher than the tabulated t value at the confidence internal, there is a statistical difference and the proposed/utilised electroanalytical protocol is not a valid procedure.

- *Case 2: Comparison of two experimental data sets: when the accepted value is unknown*

When the accepted value is unknown, a paired t-test is used to determine the validity of the experimental number. Usually, a second mean is achieved using a different instrument, another laboratory, or a secondary method within the same laboratory or a different constructed electrode.

The degree of overlap between the two distributions is evaluated, i.e. if there are significant differences between the data series at a specified level of confidence; the t-test is then:

$$\pm t = \left(\frac{(\overline{x_1} - \overline{x_2})}{s_p}\right)\sqrt{\frac{n_1 n_2}{n_1 + n_2}} \tag{B.10}$$

where $\overline{x_1}$ is the mean of one data set $\overline{x_2}$ is the mean from the second data set and s_p is called the pooled standard deviation given by:

$$s_p = \sqrt{\frac{s_1^2(n_1 - 1) + s_2^2(n_2 - 1) + \cdots s_k^2(n_k - 1)}{n_1 + n_2 \ldots n_k - k}} \tag{B.11}$$

where the value of k is the number of experimental means used for comparison. For example, if there are two sets of experimental means, then the value of k is 2.

F-test:

The F-test compares the precision of two sets of data via the following expression:

$$F = \frac{s_1^2}{s_2^2} \tag{B.12}$$

where $s_1 > s_2$. The two sets of data do not have to be obtained from the identical sample, so long as both samples are sufficiently similar such that any indeterminate errors can be considered the same. An F-test can provide insights into two main areas: (i) *Is method A more precise than method B?* and (ii) *Is there a difference in the precision of the two methods?* To calculate an F-test, the standard deviation of the method which is assumed to be more precise is placed in the denominator, while the standard deviation of the method which is assumed to be least precise is placed in the numerator. If the calculated F value is greater than a critical value

Table B.3 Selected values of F at the 95 % confidence level ($P = 0.05$)

Degrees of freedom (Denominator)	Degrees of freedom (Numerator)			
	2	3	4	5
2	19.00	19.16	19.25	19.30
3	9.55	9.28	9.12	9.01
4	6.94	6.59	6.39	6.26
5	5.79	5.41	5.19	5.05
∞	3.00	2.60	2.37	2.37

that is tabulated (Table B.3) for the chosen confidence level, then there is a significant difference at the probability level: Table B.3 shows some values for the F-test.

B.5 Recovery Tests

Such tests are performed in which a known amount of the analyte is added into a sample matrix and the chosen analysis is then performed before and after addition of the target analyte. Such recovery experiments can also be performed in control (buffers) or simulated field (in the presence of potential or known interferents) samples. The % recovery is given by:

$$\% \text{ recovery} = (Concentration found / Concentration added) \times 100 \qquad (B.13)$$

B.6 Standard Additions

Comparison against calibration standards of known analyte content allows the unknown analyte content in a sample to be deduced. When the matrix of the samples requiring measurement affects the sensitivity of the measuring instrument, standard addition calibration is typically employed. This is a widely accepted analytical technique which has been devised to overcome a particular type of matrix effect which could potentially give a biased result.

Figure B.4 shows a typical calibration plot performed in 'Matrix A' which is the matrix used for calibration, as was shown in Fig. B.1 which is typically performed in buffer solutions. Shown in Fig. B.4b is a rotational matrix effect which arises when the analytical signal is affected by non-analyte constituents of the test solution. Note that the size of such effect is proportional and hence is sometimes called a 'proportional' effect where it changes the slope of the calibration but not its intercept. Also shown in Fig. B.4a is the case of the translational effect which

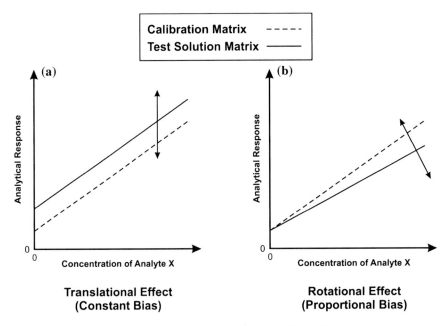

Fig. B.4 The effect of translational effects (**a**) and rotational effects (**b**) upon analytical response

arises from a signal produced by concomitant substances present in the test solution and not by the analyte, and is independent of the concentration of the analyte (often referred to as a background or baseline interference); [9, 10] note that in this case the effect is such that the intercept is affected rather than the gradient. Importantly, both interferences can have the same effect on the observed signal (point X on Fig. B.5) and thus observations at more than one concentration are required to distinguish between the two matrix effects. The inability to distinguish between the two matrix effects can give rise to misleading results. Note that the use of standard additions can correct rotational effects while translational (if present) effects have to be separately eliminated or corrected, or again the results are potentially biased.

There are two different ways to perform standard addition calibrations. The first is known as the conventional standard addition calibration (C-SAC) which compares the instrumental responses of several solutions in separate vessels containing the same quantity of sample, but different quantities of calibration standard and a blank, such that the volume in each vessel is fixed. The second is the sequential standard addition calibration (S-SAC) which compares the instrumental response from a quantity of sample in a single vessel to the instrumental response following the addition of portions of calibration solution into this same vessel, such that the volume considered in each measurement is not fixed. Quantification in both cases is performed by extrapolation of the calibration relationship produced to the intercept with the x-axis at zero analyte content,

Fig. B.5 Different types of matrix effects on the observed analytical signal. *Matrix A* is the calibration matrix, while *Matrix B* is a rotational effect which changes the size of the signal derived from the analyte but not the intercept. In the case of *Matrix C*, the intercept has been shifted by a translational effect, but the slope is unaffected. At point *X* the two matrix effects have the same outcome. Reproduced with permission from Ref. [9]

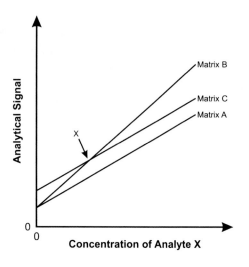

Fig. B.6 A typical standard addition plot; the analytical signal in this case is the electrochemical response. The estimated calibration line is extrapolated to zero response and the negative reading is the concentration estimate

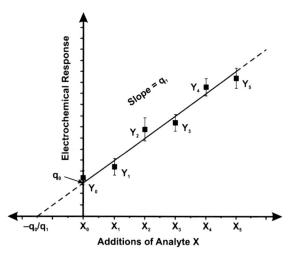

giving the content of the unknown directly (in the case of C-SAC) or indirectly (in the case of S-SAC) [9, 10, 11]. Note that the most common approach is the second approach due to its ease of experimental use. Figure B.6 shows a typical standard addition plot.

Such an approach is justified if the analyst is sure that the analytical calibration is linear over the relevant concentration range, since tests for non-linearity require a large number of measurements to have a useful degree of statistical power [9, 10].

Interestingly Thompson [9, 10] points out that one can obtain better precision in the estimation of the calibration slope if the measurements are confined to the ends of the linear range as depicted in Fig. B.7. Following several measurements of the response of the test solution and also of a single spiked solution, the best estimate of the calibration function is simply the line joining the two mean results. This

Fig. B.7 Standard additions with the spike added. This design provides a concentration c estimate with a precision slightly better than the equally spaced design (Fig. B.6). Note $N = 3$ for both "zero" and the concentration (x) of analyte added

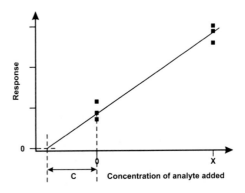

procedure gives exactly the same response function as regression, either simple or weighted. Additionally, this approach requires a smaller number of operations per measurement result.

Thompson shows that poor results, in terms of precision and relative standard deviation, will be obtained unless the concentration of the analyte is greater than about 4 times the detection limit; indeed Thompson suggested that it would pay off to always make the standard addition more than five times the analyte concentration, as long as it is consistent with the liner range of the analytical method [9, 10].

Last, Thompson's recommendations for successfully performing standard additions are as follows:

1. Make sure that the analytical method is effectively linear over the whole of the required working range.
2. Make sure that any translational interference is eliminated separately.
3. Only one level of added analyte is necessary, with repeated measurements if better precision is required.
4. Let the concentration of the added analyte be as high as is consistent with linearity, and ideally at least five times the original concentration of analyte.

It is interesting to consider who first introduced the concept of standard additions. Kelly et al. provide a useful, comical yet enlightening, overview of the origin and history of the standard addition protocol [12].

It is commonly accepted that Chow and Thompson [13] first described the standard addition protocol, and their work is frequently cited as such. However, Campbell and Carl [14] had reported the first description of the standard addition approach earlier, but their paper has received about half the citations of that of Chow and Thompson. Kelly et al. [12] suggest that both groups were working independently of each other and that the Chow and Thompson paper has been cited more since they were the first to use the standard addition method to determine the concentration of an analyte in a complex natural aqueous medium (and the protocol is the most often used with aqueous solutions). Regardless of the subtle difference between these papers, Kelly et al. [12] show that it was an

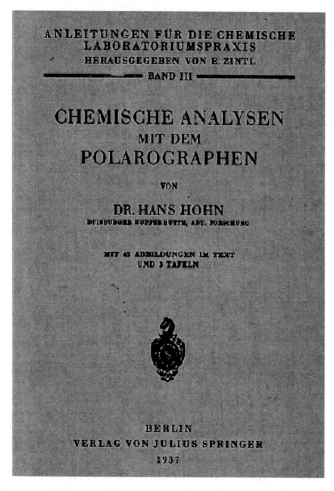

Fig. B.8 The cover of Hohn's book on polarography containing the first description of 'The Method of Standard Addition'. Reproduced with permission from Ref. [12]. Copyright Springer Science and Business Media

electrochemist who actually invented the concept of standard additions! This work was prior to that of both the papers of Chow and Thompson and Campbell and Carl.

A lesser-known person, Hans Hohn, actually reported the first description of standard additions; Fig. B.8 shows the cover of Hohn's book on polarography [13].

Hans Hohn was a mining chemist (1906–1978) and was the first to document the standard addition method in his 1937 book, which utilised polarography and contains the details of 28 experiments followed by nine detailed examples of polarographic analyses. Figure B.9 shows an original polarogram of an original

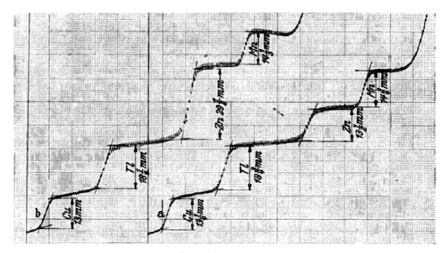

Abb. 16. Bestimmung von Zink durch Eichzusatz.
a Probe und Grundlösung, *b* dieselbe Lösung nach Eichzusatz.

Fig. B.9 Polagram of the zinc standard addition experiment originally from Ref. [15] where the figure caption reads: Determination of Zinc by calibration (standard) addition. *a* unknown and supporting electrolyte, *b* same solution after calibration addition. Reproduced with permission from Ref. [12]. Copyright Springer Science and Business Media

solution, labelled 'a', containing know amounts of Cu, Ti, Zn and Mn. Following additions of a known amount of Zn to the original solution, a second polarogram, labelled 'b' is obtained. Note that the Zn response has increased while the others are unchanged.

What follows is exceptionally intriguing. While Kelly et al. were researching who actually was the first to report standard additions, they naturally wanted to read the authoritative work of Hohn and through an inter-library loan, the authors received a copy of the book, unexpectedly from the University of Washington (this is the university where the authors Chow and Thompson reside) in Seattle. The copy of Hohn's book still retained the circulation card (this is stamped when you take a book out from the library) inside the back cover and date-due slip; this is shown in Fig. B.10.

It is clearly obvious from inspection of the circulation card (in Fig. B.10) that Professor Thompson checked-out (borrowed) Hohn's book on June 14, 1939 and was sent a notice three months later on 15 September, 1939. Later, T. J. Chow also checked-out Hohn's book on 26 November, 1951. Clearly Thompson and Chow had inspected Hohn's book prior to their publication in Analytical Chemistry [13]. Clearly Chow and Thompson re-discovered/re-reported the standard addition protocol, yet decided not to cite this seminal work. Kelly et al. carefully suggest

Fig. B.10 Circulation card and date-due slip from the copy of Hohn's book, obtained from the library of the University of Washington. It clearly shows that Professor Thomas Thompson borrowed (took-out) Hohn's book from the library on June 14, 1939 and was subsequently sent a notice three months later on 15 September, 1939. On November 26, 1951, Tsaihwa Chow also borrowed Hohn's book from the library. In 1952 they later published the standard addition method as their own, which is commonly cited in the literature as the first description of the standard addition method; see text for a full explanation. Reproduced with permission from Ref. [12]. Copyright Springer Science and Business Media

that Chow and Thompson probably thought they had extended the method of Hohn's significantly that a citation was not necessary. Whatever the reason, the story is highly fascinating.

Appendix C
Experimental Tips for the Graphene Electrochemist

Below is a non-exhaustive list of tips and considerations that a 'Graphene Electrochemist' should undertake during the course of their work.

- Characterise your fabricated graphene appropriately, especially to confirm that the material is indeed corresponding to the physicochemical characteristics of graphene (TEM and Raman spectroscopy for instance), along with the reporting the O/C ratio (XPS for example) and the presence (% quantity) or absence of surface defects if possible.
- Undertake appropriate control experiments using other carbon allotropes, such as graphite, carbon black and amorphous carbon (as explored in energy related areas). See for example Refs: [16–19]. The direct comparison of graphene modified electrodes with edge plane and basal plane pyrolytic graphite electrodes (EPPG and BPPG respectively) is particularly essential in order to justify the benefits of graphene modified electrodes [16–18].
- Consider the effect of surface coverage at modified electrodes; three zones (*zone I, II and III*) can be encountered (see Chap. 3 and Ref. [17]) and it is important to deduce which one you are in.
- Explore the effect of the underlying/supporting electrode substrate; this will affect how the graphene aligns (i.e. its orientation; also consider this factor separately) and dictate the observed electrochemical response; see Chap. 3 [17].
- Ensure that improvements observed in peak potentials (electron transfer kinetics) from using graphene are not simply due to changes in mass-transport, that is, giving rise to thin-layer type behaviour (*zone III* – see Chaps. 2 and 3) [16, 17].
- Perform control experiments with surfactant modified electrodes if your graphene has been fabricated in the presence of such surfactants, or where they are used to reduce coalescing [16, 20–22]. These controls should extend to the solvents utilised where graphene is suspended in solution [16, 17, 22].

D. A. C. Brownson and C. E. Banks, *The Handbook of Graphene Electrochemistry*, DOI: 10.1007/978-1-4471-6428-9, © Springer-Verlag London Ltd. 2014

- Consider the effect of impurities (metallic, carbonaceous, etc.) on the electrochemical performance and perform the appropriate control experimental [23, 24]. For example, metallic impurities might arise from using an impure graphite starting material for producing graphene oxide and graphene from reduced graphene oxide—such impurities can come from using low grade acid that contains metals [25].
- Consider the effect that oxygenated species have on the electrochemical response; the fabricated graphene may have specific surface groups present that give rise to beneficial responses, rather than that of true graphene [16, 18].
- If using graphene for energy storage applications try and ensure that the selected current densities (A/g) are the same as literature reports, this allows a direct comparison of the energy storage (F/g) capabilities [26].
- In the case of CVD grown graphene that remains supported on the metal catalyst (or in other instances where the graphene of interest is deposited onto an electroactive support surface, as to electrically wire/connect to the graphene), ensure that the underlying metal surface (usually nickel or copper) is not accessible to the solution as this can give rise to misinterpretations of the experimental data. Control experiments using the bare metal surface (or unmodified alternative support surface) with the electrochemical species under investigation should be performed [18, 23, 27].

Appendices References

1. Web-Resource, The 2010 nobel prize in physics—press release, Nobelprize.org (2012), http://www.nobelprize.org/nobel_prizes/physics/laureates/2010/press.html. Accessed 28 Feb 2012
2. Web-Resource, Graphene—letter from Walt de Heer, Georgia Institute of Technology (2012), http://www.gatech.edu/graphene/. Accessed 30 July 2012
3. IUPAC, *Compendium of chemical terminology*, 2nd edn. (the "Gold Book"). Compiled by A. D. McNaught, A. Wilkinson (Blackwell Scientific Publications, Oxford, 1997), XML on-line corrected version: http://goldbook.iupac.org (2006) created by M. Nic, J. Jirat, B. Kosata, updates compiled by A. Jenkins, ISBN 0-9678550-9-8. doi:10.1351/goldbook
4. J. Mocak, A.M. Bond, S. Mitchell, G. Scollary, Pure Appl. Chem. **69**, 297–328 (1997)
5. M. Thompson, S.L.R. Ellison, R. Wood, Pure Appl. Chem. **74**, 835–855 (2002)
6. J. Mocak, I. Janiga, E. Rabarova, Nova Biotechnologica **9**, 91–100 (2009)
7. R.B. Dean, W.J. Dixon, Anal. Chem. **23**, 636–638 (1951)
8. D.B. Rorabacher, Anal. Chem. **63**, 139–146 (1991)
9. AMC Technical Briefs, Royal Society of Chemistry: Standard Additions: Myth and Reality, 2009, No. 37. Editor: Michael Thompson. ISSN 1757-5958
10. S.L.R. Ellison, M. Thompson, Analyst **133**, 992–997 (2008)
11. R.J.C. Brown, T.P.S. Gillam, Measurement **45**, 1373–1670 (2012)
12. W.R. Kelly, K.W. Pratt, W.F. Guthrie, K.R. Martin, Anal. Bioanal. Chem. **400**, 1805–1812 (2011)
13. T.J. Chow, T.G. Thompson, Anal. Chem. **27**, 18–21 (1955)
14. W.J. Campbell, H.F. Carl, Anal. Chem. **26**, 800–805 (1954)
15. H. Hohn, *Chemische Analysen mit dem Polarograhen* (Springer, Berlin, 1937)
16. D.A.C. Brownson, C.E. Banks, Analyst **135**, 2768–2778 (2010)
17. D.A.C. Brownson, L.J. Munro, D.K. Kampouris, C.E. Banks, RSC Adv. **1**, 978–988 (2011)
18. D.A.C. Brownson, C.E. Banks, Phys. Chem. Chem. Phys. **14**, 8264–8281 (2012)
19. C.H.A. Wong, A. Ambrosi, M. Pumera, Nanoscale **4**, 4972–4977 (2012)
20. D.A.C. Brownson, J.P. Metters, D.K. Kampouris, C.E. Banks, Electroanalysis **23**, 894–899 (2011)
21. D.A.C. Brownson, C.E. Banks, Electrochem. Commun. **13**, 111–113 (2011)
22. D.A.C. Brownson, C.E. Banks, Chem. Commun. **48**, 1425–1427 (2012)
23. D.A.C. Brownson, C.E. Banks, Phys. Chem. Chem. Phys. **13**, 15825–15828 (2011)
24. D.A.C. Brownson, M. Gomez-Mingot, C.E. Banks, Phys. Chem. Chem. Phys. **13**, 20284–20288 (2011)
25. A. Ambrosi, C.K. Chua, B. Khezri, Z. Sofer, R.D. Webster, M. Pumera, Proc. Natl. Acad. Sci. U. S. A. **109**, 12899–12904 (2012)
26. D.A.C. Brownson, D.K. Kampouris, C.E. Banks, J. Power Sources **196**, 4873–4885 (2011)
27. D.A.C. Brownson, C.E. Banks, RSC Adv. **2**, 5385–5389 (2012)

D. A. C. Brownson and C. E. Banks, *The Handbook of Graphene Electrochemistry*, 201
DOI: 10.1007/978-1-4471-6428-9, © Springer-Verlag London Ltd. 2014